A Beginner's Book of TEX

Raymond Seroul
Silvio Levy

A Beginner's Book of TEX

Foreword by Dominique Foata

Springer-Verlag
New York Berlin Heidelberg London
Paris Tokyo Hong Kong Barcelona

Raymond Seroul
Université Louis Pasteur
Laboratoire de Typographie Informatique
7, rue Rene-Descartes
67084 Strasbourg, France

Silvio Levy
Geometry Center
University of Minnesota
1300 South Second Street
Minneapolis, MN 55415 USA

Cover mathematics adapted from A.G. Alings, *Superficierum curvatura* (dissertation), Groningen (1849).

Translated and adapted by Silvio Levy from the original French *Le petit livre de TEX*, by Raymond Seroul, © 1989, InterEditions, Paris.

Quotation on pages 10–11 from *The Hobbit*, by J.R.R. Tolkien, © 1979, Allen and Unwin, pp. 16–17 (paperback edition, 1979).

Library of Congress Cataloging-in-Publication Data
Seroul, Raymond.
    A beginner's book of TeX / Raymond Seroul, Silvio Levy.
        p.   cm.
    Includes bibliographical references and index.
    ISBN 0-387-97562-4. — ISBN 3-540-97562-4 (Berlin)
    1. TeX (Computer systems)   2. Computerized typesetting.
    3. Mathematics printing.   I. Levy, Silvio Vieira Ferreira.
    II. Title.
    Z253.4.T47S47 1991
    686.2'2544—dc20   91-19278

Printed on acid-free paper.

Photocomposed copy prepared from author's TEX file.
Printed and bound by R.R. Donnelley & Sons, Harrisonburg, VA.
Printed in the United States of America.

9 8 7 6 5 4 3 2 1

ISBN 0-387-97562-4 Springer-Verlag New York Berlin Heidelberg
ISBN 3-540-97562-4 Springer-Verlag Berlin Heidelberg New York

Foreword

The last two decades have witnessed a revolution in the realm of typography, with the virtual disappearance of hot-lead typesetting in favor of the so-called digital typesetting. The principle behind the new technology is simple: imagine a very fine mesh superimposed on a sheet of paper. Digital typesetting consists in darkening the appropriate pixels (tiny squares) of this mesh, in patterns corresponding to each character and symbol of the text being set. The actual darkening is done by some printing device, say a laser printer or phototypesetter, which must be told exactly where the ink should go.

Since the mesh is very fine—the dashes surrounding this sentence are some six pixels thick, and more than 200 pixels long—the printer can only be controlled by a computer program, which takes a "high-level" description of the page in terms of text, fonts, and formatting commands, and digests all of that into "low-level" commands for the printer. TEX is such a program, created by Donald E. Knuth, a computer scientist at Stanford University.

Knuth distilled into his program generations of typesetting wisdom, and as a consequence it is easy to produce in TEX documents having a highly professional appearance. Authors in the scientific and technical world quickly learned to use TEX for their preprints and reports, only to see them entirely reset before publication, often with inferior results. Indeed, the printing establishment had become computerized, but the software used was a long way from producing results of the quality that seasoned professionals could achieve with the old technology.

Lately, publishers and printers have overcome their skepticism that an author or technical typist, aided by a public-domain program born in academia, might generate beautiful documents worthy of the most venerable typographical tradition. Many publishers now accept diskettes and tapes with TEX files, and send them to the printer after making only minimal changes, to ensure conformity of style and correct treatment of complex material such as displays and tables. Other publishers have trained in-house staff to rekey manuscripts in TEX. The technological revolution is complete!

Authors and typists are faced then with the same task: learning T$_E$X. Some will be content with picking up the basics, and apply them to get output that already looks surprisingly good. Others will go on to assimilate advanced techniques, and perform typographical *tours de force*. Either way, one must start with the fundamentals. Knuth's manual and system description, *The T$_E$X book*, is a superbly written reference, but is hard for a beginner to absorb. There is still a lack of introductory books to help potential users get started on their own and quickly.

The present book tries to answer this need. It contains a careful explanation of all fundamental concepts and commands, but also a wealth of commented examples and "tricks," based on the authors' long experience with T$_E$X. The attentive reader will quickly be able to create a table, or customize the appearance of the page, or code even the most complicated formula. The last third of the book is devoted to a Dictionary–Index, summarizing all the material in the text and going into greater depth in many areas.

Dominique Foata
Director, Laboratoire de Typographie Informatique
Université Louis Pasteur, Strasbourg, France

Contents

Chapter 4: The fonts TₑX uses   27

Chapter 5: Spacing, glue and springs   38

Chapter 6: Paragraphs   52

Chapter 7: Page layout   64

Introduction

This book was born from the first author's desire to supply the French-speaking community with a readily accessible introduction to TEX, at a time when there was no such thing even in English, and no TEX manual in French at all. The success enjoyed by *Le Petit Livre de TEX* caused Springer to commission its translation into English. This turned out to be no straightforward task; many sections were adapted and revised by the translator-turned-coauthor, and new material added, especially to Chapters 12 and 13.

This work is addressed primarily to beginner and intermediate users. Everywhere we have tried to put ourselves in the beginner's shoes and ask, What would have made this topic clearer when we first learned it? At the same time, we have tried to keep in mind the diversity of backgrounds that characterizes TEX users: technical typists, authors in the sciences, in math, in engineering. TEX is also making inroads in the humanities, thanks to its capable handling of footnotes, bibliographies, indexes, accents...

If you are a beginner, read each chapter selectively, skipping whatever appears too technical. Concentrate on Chapters 1–7, and also on Chapter 11 if you have to type mathematics. (This chapter is long but mostly very easy.) Don't worry about understanding everything. The best way to learn TEX is by example, and this book is full of them. Copy these examples and modify them, experimenting with everything that looks like it can be changed. You will be learning by osmosis, without pain. There's no substitute for experimentation!

Very soon you will find yourself no longer a beginner. You'll be wanting to tinker with the page layout, put boxes together, use a variety of fonts, make tables, define new commands. Then it is time to read the relevant chapters more systematically.

After you have assimilated the material in Chapters 1–12, you can consider yourself an experienced user—a TEX "master." But we hope this book will remain useful: that's why we have wrapped it up with an extensive Dictionary and Index of TEX commands and concepts. In addition to repeating much of the information from

Chapters 1–12, for ease of reference, the Dictionary treats many concepts in greater depth, and includes some commands that are not mentioned elsewhere, but are likely to be useful to an aspiring "wizard." Beyond that, you will have to refer to *The TEXbook*, which remains the definitive reference for TEX.

We encourage you to join the TEX Users Group, which entitles you to a subscription to *TUGboat*, a journal containing news, tutorials, program listings, conference announcements, advertisements, etc. The TUG office itself is a primary source of information on TEX problems; when the staff does not know the answer to a question, it can generally put you in touch with someone who does. The address is P.O. Box 9506, Providence, RI 02940.

We would like to thank all those who helped us in our learning of TEX, and also those who have come to us with TEX questions. This unending stream of questions has increased our experience tenfold.

We would appreciate any reader feedback for future editions. Corrections of typos or mistakes that may have eluded us are also welcome.

Raymond Seroul
Silvio Levy

1
What is T_EX?

1.1 The birth of T_EX

T_EX was created by Donald E. Knuth, a professor at Stanford University who has achieved international renown as a mathematician and computer scientist. Knuth also has an aesthetic sense uncommon in his field, and his work output is truly phenomenal.[1]

T_EX is a happy byproduct of Knuth's mammoth enterprise, *The Art of Computer Programming*. This series of reference books, designed to cover the whole gamut of programming concepts and techniques, is a *sine qua non* for all computer scientists; three volumes were published in the seventies, and, after a long hiatus during which Knuth devoted himself to computer typesetting, a fourth is about to follow suit. Here is, in Knuth's own words, the story of T_EX's birth:[2]

> Why did I *start* working on T_EX in 1977? The whole thing actually began long before, in connection with my books *The Art of Computer Programming*. I had prepared a second edition of volume 2, but when I received galley proofs they looked awful—because printing technology had changed drastically since the first edition had been published. The books were now done with phototypesetting, instead of hot lead Monotype machines; and

---

[1] These comments are not exaggerated. We know his books well, have studied some of his theoretical articles, and have used T_EX for quite a while now. He is simply amazing!

The T_EXbook, published by Addison-Wesley, was written and typeset—in T_EX, needless to say—by Knuth himself. One can't imagine a better introduction to this man's multifaceted talent.

[2] Excerpted from "Remarks to celebrate the publication of *Computers and Typesetting*," address delivered at the Computer Museum in Boston on May 21, 1986. The full text can be found in *TUGboat*, **7** (1986), 95–98. (*TUGboat* is the T_EX Users Group newsletter; for more information, see the Appendix.)

(alas!) they were being done with the help of computers instead of by hand. The result was poor spacing, especially in the math, and the fonts of type were terrible by comparison with the original. I was quite discouraged by this, and didn't know what to do. Addison-Wesley offered to reset everything by the old Monotype method, but I knew that the old way was dying out fast; surely by the time I had finished volume 4 the same problem would arise again, and I didn't want to write a book that would come out looking like the recent galleys I had seen.

Then ... we received galley proofs of [Pat Winston's *Artificial Intelligence*, which] had been made on a new machine in Southern California, all based on a discrete high-resolution raster ... The digital type looked a lot better than what I had been getting in my own galley proof ... Within a week after seeing the galley of Winston's book, I decided to drop everything else and work on digital typography ...

Ever since these beginnings in 1977, the TEX research project that I embarked on was driven by two major goals. The first goal was *quality*: we wanted to produce documents that were not just nice, but actually the best ... By 1977 there were several systems that could produce very attractive documents. My goal was to take the last step and go all the way, to the finest quality that had ever been achieved in printed documents.

It turned out that it was not hard to achieve this level of quality with respect to the formatting of text, after about two years of work. For example, we did experiments with *Time* magazine to prove that *Time* would look much better if it had been done with TEX. But it turned out that the design of typefaces was much more difficult that I had anticipated; seven years went by before I was able to generate letterforms that I began to like.

The second major goal was to be *archival*: to create systems that would be independent of changes in printing technology as much as possible. When the next generations of printing devices came along, I wanted to be able to retain the same quality already achieved, instead of having to solve all the problems anew. I wanted to design something that would still be usable in 100 years. In other words, my goal was to arrange things so that, if book specifications are saved now, our descendants should be able to produce an equivalent book in the year 2086 ...

1.2 How TEX works

Roughly speaking, text processors fall into two categories:

• WYSIWYG systems: what you see is what you get.[3] You see on the screen at all times what the printed document will look like, and what you type has immediate effect on the appearance of the document.

---

[3] This slogan is 100% true only if screen and printer use the same resolution and page description language.

- *markup* systems, where you type your text interspersed with formatting instructions, but don't see their effect right away. You must run a program to examine the resulting image, whether on paper or on the screen. In computer science jargon, markup systems must *compile* the *source file* you type.

WYSIWYG systems have the obvious advantage of immediate feedback, but they are not very precise: what is acceptable at a resolution of 300 dots per inch, for an ephemeral publication such as a newsletter or flier, is no longer so for a book that will be phototypeset at high resolution. The human eye is extraordinarily sensitive: you can be bothered by the appearance of a text without being able to pinpoint why, just as you can tell when someone plays the wrong note in an orchestra, without being able to identify the culprit. One quickly learns in typesetting that the beauty, legibility and comfortable reading of a text depend on minute details: each element must be placed exactly right, within thousandths of an inch. For this type of work, the advantage of immediate feedback vanishes: fine details of spacing, alignment, and so on are much too small to be discernible at the screen's relatively low resolution, and even if it such were not the case, it would still be a monumental chore to find the right place for everything by hand.

For this reason it is not surprising that in the world of professional typesetting markup systems are preferred. They automate the task of finding the right place for each character with great precision. Naturally, this approach is less attractive for beginners, since one can't see the results as one types, and must develop a feeling for what the system will do. But nowadays, you can have the best of both worlds by using a markup system with a WYSIWYG *front end*; we'll talk about such front ends for T_EX later on.

T_EX was developed in the late seventies and early eighties, before WYSIWYG systems were widespread. But were it to be redesigned now, it would still be a markup language. To give you an idea of the precision with which T_EX operates: the internal unit it uses for its calculations is about a hundred times smaller than the wavelength of visible light! (That's right, a hundred times.) In other words, any round-off error introduced in the calculations is invisible to the naked eye.

The result of T_EX's lucubrations is not the complete image of a printed page, but rather an abstract description of it. This description is independent of the machine where you ran T_EX, and of the printer that will create the hard copy—in other words, it is completely *portable*. Here is the decoded version of a tiny portion of a page, containing the T_EX logo:

```
level 1:(h=0,v=655360,w=0,x=0,y=0,z=0,hh=0,vv=42)
109: fntdef1 0: cmr10---loaded at size 655360 DVI units
130: fntnum0 current font is cmr10
[T]
level 2:(h=1784036,v=655360,w=0,x=0,y=0,z=0,hh=113,vv=42)
137: down3 141084 v:=655360+141084=796444, vv:=51
[E]
level 2:(h=1784036,v=655360,w=0,x=0,y=0,z=0,hh=113,vv=42)
[ X]
```

1.3 The good news and bad news about TEX

The good news

First of all, TEX produces documents of unusually *high quality*, especially in the case of math. As we've mentioned, Knuth is an aesthete, and he made a point of incorporating in his program all the wisdom of generations of typographers. Here are some examples of the extraordinary care with which TEX treats your text:

• It handles ligatures automatically, in the best typesetting tradition. It also does automatic kerning, that is, it sets characters whose shapes "match" closer together, so the spacing between characters looks uniform.

• It has an intricate mechanism for justifying lines, resorting when necessary to hyphenation. The hyphenation rules themselves can be reconfigured, so as to adapt TEX to different languages.

• The spacing between the various components of a mathematical formula is determined by TEX according to traditional rules used by the best math typesetting houses. It is very rare that a formula comes out looking "wrong". And typing a mathematical formula in TEX is so easy, natural and logical that one finds oneself doing it just for the heck of it...

A well-written TEX document is formatted by means of *macros*, that indicate how each component should be typeset. A macro is a short program that saves you from having to give explicit formatting instructions. For instance, \footnote lets you include a footnote without worrying about moving the text to the bottom of the page or typing little numbers above the line. Macros are written in terms of *primitives*, like \indent or \par (for paragraph), which form the basic vocabulary of TEX.

Macros make TEX immensely *versatile*. To modify the appearance of a document, it is enough to change the definition of certain macros, *without touching the text*. There are efforts underway to define standards for the coding of the structure of on-line documents. TEX will fit right in with these standards, since it can be used as a high-level document description language.

TEX is *portable*. A document written in TEX, containing your texts and macros for formatting, can be coded entirely using characters from the printable ASCII set, in the range 32–126 (plus the carriage return), *even if it prints characters in foreign alphabets that have higher codes*. If this is Greek to you, here's what it means in practice: to share a text with a friend anywhere in the world, turn on your modem, send your file, and presto! No need for special encodings, conversions or anything.[4] With computer networks spanning the whole globe, the possibility of sending formatted texts through them is an obvious advantage. And even the most complicated scientific text can be written in TEX, using only ASCII characters.

To give just one example, the database maintained by the American Mathematical Society, or AMS, is based on TEX. It contains abstracts of all mathematical articles

---

[4] Ideally, that is. In practice, some characters can get mangled when they go through certain networks: the backslash is especially susceptible. A good trick is to list all the ASCII characters at the top of your file, in order, so the recipient can at least tell what's going on. At any rate, the problem seems to be increasingly rarer nowadays.

published in the world. By accessing the database you can obtain the abstract of any article that interests you, and read it either in T_EX source form, or formatted, after running it through T_EX.

T_EX is also portable across computers, because it doesn't depend on the peculiarities of each computer's character set, and because its calculations are done in a completely machine-independent way. A text written in T_EX looks the same (disregarding variations in printer quality) whether run on a Macintosh, a PC clone, a UNIX workstation, an IBM mainframe or even a Cray. We've tried it out: an article, written on a PC clone at the University of Strasbourg, was sent to the United States to be phototypeset at the AMS. It came out without any problems, and looked just the same as our proofs run on a humble dot-matrix printer.

T_EX doesn't create an image, just a page description. To print your document, you take T_EX's output and give it to a *driver* program, capable of transforming this description into commands that the printer can understand. In this way, T_EX is also independent of the technology of printers; when the technology changes, it's enough to write a new driver, a relatively simple program.

T_EX is much more than a text processor—it's a programming language! It is easily adaptable to your needs. You can create new commands or modify T_EX's behavior by changing its variables. With more experience, you can define new styles and write sophisticated macros for special purposes—or you can copy them from someone else. Because T_EX is portable and widespread, most things you're likely to want to do have already been done by someone else, and it's a matter of finding it. This is not always easy, of course, but a good place to start is *TUGboat*, the T_EX Users Group newsletter.

T_EX is also *extensible*—as we've seen, Knuth had an eye on the future when he created T_EX. For this reason he structured it in layers, like an onion: at the center are T_EX's 300 or so primitives, the building blocks of T_EX. Primitives, as their name implies, are very "primitive"—you wouldn't want to use them all the time. Next come higher-level commands, or macros, defined in a *format file*. The most common format file is called `plain.tex`, and it defines about 600 commands. (No need to panic! You'll need to know less than a hundred to format even fairly complicated documents, and they mostly have very natural names.) The combination of primitives and commands defined in `plain.tex` is generally called plain T_EX. On top of that you can use one of several *packages*; they provide even higher-level commands, such as `\chapter` or `\theorem`, leaving all the formatting to the system.[5] And finally, you can add your own commands. Once you become intimately familiar with T_EX, you might even write your own formats, to complement or replace plain T_EX.

T_EX is very *well-debugged*. Of course, like any program, it will never be bug-free; but since Knuth offers a prize for each new bug reported, there is an army of bug hunters out there that has sifted through every line of the code. Any remaining

---

[5] Packages are generally combined with the underlying format, giving rise to different "avatars" of T_EX: L^AT_EX, A_MS-T_EX, and so on. L^AT_EX is probably the best-known and the most complete.

bugs must be extremely recondite and unlikely to occur spontaneously. If you find one, you'll earn your prize and a place in the official listing of TEX's (former) bugs, periodically published in *TUGboat*.

TEX is *in the public domain*: Knuth shared it freely with the world. You can copy the source of TEX from anyone. When you buy TEX, you're paying solely for its implementation on a particular machine, and for a support environment, typically consisting of a driver, previewer, text editor, and so on. Proprietary systems comparable with TEX sell for ten or twenty times as much.

The TEX logo and the copyright of *The TEXbook* belong to the AMS, which is in charge of maintaining the TEX standard. For a new implementation to have the right to be called TEX it must pass a so-called "torture test," designed by Knuth himself and perfected every year.

The Pascal source of TEX, with full explanations, has been published as volume B of *Computers and Typesetting* (there are five volumes; volume A is *The TEXbook*). This in itself is remarkable: not many program sources are made into books! If you have a chance, take a look at *TEX: The Program*. You'll see how Knuth, once again, innovated: instead of presenting a dry listing, he weaves code and commentary in a beautifully typeset document.

The bad news

We now come to the shortcomings of TEX. As you will see, most of them can be and have been circumvented, usually by means of extensions or supporting programs. These are implementation-dependent and not really part of TEX, yet it is due to TEX's robustness of design that it is even possible to extend it in so many directions.

TEX programming is subtle and takes time to master. Don't worry: this is not a problem for the ordinary user. Using a macro, or even defining a simple macro, is no harder than tuning a radio. But writing a complicated macro, or designing a package, is more like putting the radio together: not a task for beginners.

TEX has a limited amount of memory, fixed for each implementation—it doesn't grow dynamically. On computers with at least four megabytes of memory you can run a version of TEX, written in C, whose memory limits are generous enough to be considered irrelevant; but on anything smaller you can run into trouble if you don't take certain precautions.

TEX uses its own fonts, which must be kept around. This is not really a design limitation, because in fact TEX can use any font whose metric information is known (see section 1.5); but until recently this information was difficult to obtain for non-TEX fonts. Nowadays many installations of TEX, both on PC's and on bigger systems, can handle PostScript fonts. With the ever-increasing diffusion of TEX, especially on small computers, it is likely that the number of available fonts will grow very quickly.

TEX can't handle slanted lines or any other graphics. There are macro packages that define simple graphics commands, but they tend to use large amounts of memory. A more promising approach is based on an escape hatch that Knuth built into TEX, foreseeing exactly this kind of situation: the \special command. This command lets you sneak into TₑX's output anything that is of no use to TₑX, but can be meaningful to the driver: for example, the name of a file containing a figure, or even raw PostScript commands. Needless to say, anything like that is highly implementation-dependent.

TₑX is not interactive. This is probably the one most common criticism of TₑX. As we mentioned in the previous section, there are reasons for that; there are also ways to get around it. Many user-friendly front ends for TₑX are now available; on the Macintosh, for example, a program called TₑXtures offers a very attractive interface. From within the same program you have access to:

- a multi-window text editor;
- TₑX proper;
- a previewer, that shows on the screen what the output will look like;
- a printer driver.

And you can even insert into your text PostScript images generated by other programs.

On PC clones, you can only run one program at a time: you must first edit a file, then run it through TₑX, then use a previewer to look at your output on the screen. You must then make a note of all the mistakes, quit the previewer and edit your file again to correct them. And so on, until the process converges. Oh, for a mouse and window interface!

As microcomputers become more powerful, the last three shortcomings we've discussed will tend to disappear. Eventually we'll have, even on microcomputers, very user-friendly systems of the type that is already available on workstations (see the Appendix). Such systems use a TₑX "engine," but they work essentially as if they were WYSIWYG: you can build or change a math formula with the mouse, without having to edit the TₑX source, and without having to know much about TₑX at all!

1.4 TₑX: who and what for?

TₑX is not a text processor. It was designed with a precise goal in mind: writing scientific texts. Scientific texts are, from a typographic point of view, paradoxical: they can contain unbelievably hairy formulas, but in terms of page layout they are generally very simple, just a series of rectangles—the paragraphs—stacked one atop another. For this reason, TₑX is unbeatable for typesetting math, or scientific copy in general; but it sputters and chokes if you give it, say, a newsletter or a complicated page layout.

TₑX was designed for scientists by a scientist (and aesthete). But certain of its features will also interest those in the humanities: accents in foreign languages, footnotes, indexing, adaptable hyphenation, programming capabilities. Knuth himself has proposed an extension of TₑX (not to be called TₑX, to maintain standardization) which can typeset copy containing both left-to-right text, as in English, and right-to-left, as in Arabic or Hebrew.

To sum up, Knuth made good his word. The scientific community now possesses a professional typesetting tool of very high level, at a price within anyone's reach. As technology evolves, we expect to see continual evolution in interfaces for TEX, while the TEX kernel will stay the same, since it is machine- and implementation-independent. The excruciating precision of TEX's internal calculations guarantees that a book done in TEX today will be printable in a hundred years without modification.

TEX is the highest-quality scientific typesetting program currently available on microcomputers. Of course, it is also available on workstations and bigger machines.

A last "argument" in favor of TEX: experience has shown that writing a scientific typesetting program is a monumental task. Since the market for such systems is not huge, it will be some time before someone succeeds in supplanting Knuth's work.[6]

1.5 TEX processing: an overview

Exactly what you commands you type or what buttons you click in order to process a TEX document depends on what system you're working on, but there are always three steps involved. This section explains what the steps mean, but you'll have to refer to the documentation that came with your TEX implementation for the details, or else ask around.

Step 1: preparing the source

Suppose you want to use TEX to typeset a letter, or an article, or a book. The first step is to type the text into a file on your computer disk, using a text editor. Together with the text you will probably want to include TEX formatting commands, or control sequences:

You should be sure to understand the difference between a text editor and a text processor. A text processor is a text editor together with formatting software that allows you to switch fonts, do double columns, indent, and so on. A text editor puts your text in a file on disk, and displays a portion of it on the screen. It doesn't format your text at all.

We insist on the difference because those accustomed to WYSIWYG systems are often not aware of it: they only know text processors. Where can you find a text editor? Just about everywhere. Every text processor includes a text editor which you can use. But if you use your text processor as a text editor, be sure to save your file using a "save " or "save text only" option, so that the text's processor's own formatting commands are stripped off. If you give TEX a file created without this precaution, you'll get garbage, because TEX cannot digest your text processor's commands.

---

[6] From this point of view, there is an interesting parallel between TEX and Fortran . . .

Step 2: Running TEX proper

TEX's actions can be schematically represented like this:

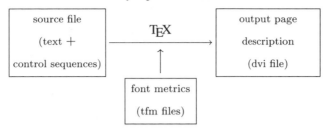

When TEX sets your text, its first task is to replace the characters in the text by their dimensions. A character has three associated dimensions: height, depth and width. Of course, these dimensions depend on the font you're using: an 'a' has different widths depending on whether it comes from a roman or a boldface font. The dimensions of characters are contained in special files, called `tfm` files, for *TEX font metrics*. The same files contain other tidbits of information, concerning ligatures (automatically managed by TEX) and italic corrections (a tiny bit of space that you can leave after an italicized word so it will look better).

After it's read a whole paragraph and converted it into these integer dimensions, TEX adds, subtracts, multiplies and divides these numbers at full throttle, and comes up with an abstract description of what the paragraph will look like on the page. This description is written into a file called the `dvi` file (for *device independent*).

Step 3: Getting output

TEX only works with dimensions: it completely ignores the shapes of the characters. When the `dvi` file is completed, another program must take over to actually produce a page: a *driver*, which sends the page to a printer, or a *previewer*, which displays it on your computer screen.

Here's what happens then:

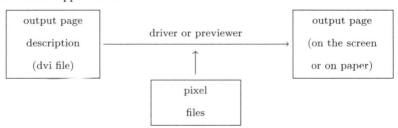

The driver (or previewer) looks for the shape of the characters in another type of font files, called *pixel files*. With this information it proceeds to create the right pattern of pixels on your screen or on a sheet of paper.

This overview of the way TEX works explains why you can't just use any font with TEX: you must have the metric information in a `tfm` file. More and more fonts nowadays come with `tfm` files. There are also programs that create `tfm` files for PostScript fonts, starting with the metric files (`afm` files) provided with such fonts.

1.6 Looking ahead

We have just discussed in detail the qualities and shortcomings of TEX. It is now time for you to try your hand at it. To do this, you should input one or both of the two short texts below on your computer. (Skip the second if you're not interested in typesetting math.)

To begin with, compare the source with the page output. Try to guess the result of each control sequence (word starting with a backslash). Then use your favorite text editor to type the first text into a file `hobbit.tex`, and the second into a file `math.tex`.

TEX differs from most WYSIWYG systems in its treatment of the two "invisible" characters, the space and the carriage return, which we will generally write SP and CR. (The carriage return is what you type to start a new line on the screen.) Normally, when you use TEX, you can start a new line whenever you want: you can even type a single word per line, and the result will be the same, because TEX justifies output lines, making them all the same length. For TEX, a CR has the same effect as an SP. Several SP in a row also have the same effect as a single one. But for these two texts, it's best if you try to input the lines exactly as shown, so it'll be easier to compare your file with the model if you type something wrong.

There is an exception to the rule that a CR is the same as an SP: when TEX sees two CR in a row—which is to say, when it sees an empty line—it starts a new paragraph. So you should also respect empty lines when typing in these texts.

Once you've typed in one or both files, you should run TEX on them and send the output to the printer. In almost all implementations, the command to run TEX on a file called `hobbit.tex` is `tex hobbit`. The command to print the resulting `dvi` file, which contains the page description, is completely system-dependent, and you'll have to consult the documentation or ask someone around to find out what to do.

While you're running TEX you may run into error messages, indicating that you made typos when inputting the files. Respond to each error message with a CR, then edit the source file again to make sure it follows exactly the model on the previous pages. Eventually TEX will run without any error messages, and you will be able to send the `dvi` file to the printer and look at the output.

First text

```
\hsize=115mm

''Good morning!'' he said at last. ''We don't want any
adventures here, thank you! You might try over The Hill
or across The Water.'' By this he meant that the
conversation was at an end.

''What a lot of things you do use {\it Good morning\/}
for!'' said Gandalf. ''Now you mean that you want to get
rid of me, and that it won't be good till I move off.
```

```
''Not at all, not at all, my dear sir! Let me see,
I don't think I know your name?''

''Yes, yes, my dear sir!---and I do know your name,
Mr.~Bilbo Baggins. And you do know my name, though you
don't remember that I belong to it. I am Gandalf, and
Gandalf means me! To think that I should have lived
to be good-morninged by Beladonna Took's son, as if I was
selling buttons at the door!''

''Gandalf, Gandalf! Good gracious me! Not the wandering
wizard that gave Old Took a pair of magic diamond studs
that fastened themselves and never came undone till ordered?
Not the fellow who used to tell such wonderful tales at
parties, about dragons and goblins and giants and the
rescue of princesses and the unexpected luck of widows'
sons? \dots\ Bless me, life used to be quite inter---I mean,
you used to upset things badly in these parts once upon
a time. I beg your pardon, but I had no idea you were
still in business.''\footnote*{J. R. R. Tolkien,
{\it The Hobbit.}}

\bye
```

"Good morning!" he said at last. "We don't want any adventures here, thank you! You might try over The Hill or across The Water." By this he meant that the conversation was at an end.

"What a lot of things you do use *Good morning* for!" said Gandalf. "Now you mean that you want to get rid of me, and that it won't be good till I move off.

"Not at all, not at all, my dear sir! Let me see, I don't think I know your name?"

"Yes, yes, my dear sir!—and I do know your name, Mr. Bilbo Baggins. And you do know my name, though you don't remember that I belong to it. I am Gandalf, and Gandalf means me! To think that I should have lived to be good-morninged by Beladonna Took's son, as if I was selling buttons at the door!"

"Gandalf, Gandalf! Good gracious me! Not the wandering wizard that gave Old Took a pair of magic diamond studs that fastened themselves and never came undone till ordered? Not the fellow who used to tell such wonderful tales at parties, about dragons and goblins and giants and the rescue of princesses and the unexpected luck of widows' sons? ... Bless me, life used to be quite inter—I mean, you used to upset things badly in these parts once upon a time. I beg your pardon, but I had no idea you were still in business."*

---

* J. R. R. Tolkien, *The Hobbit.*

Second text

```
\hsize=115mm

\centerline{\bf 3. Endomorphisms of an $A$-module
of finite type}

\medskip
\noindent Theorem 3.1.
{\it If $M$ is an $A$-module of finite type and
$u:M\rightarrow M$ is an endomorphism of $M$, the following
equivalence holds:
$$
u \hbox{ is surjective }\iff u \hbox{ is bijective.}
$$}% end of italics

The direction $\Leftarrow$ is obvious. We show
the opposite direction $\Rightarrow$. Let
$(x_1,x_2,\ldots,x_n)$ be generators for $M$.
Since $u$ is surjective, there exist $y_i\in M$ such
that $x_i=u(y_i)$. Since the $x_i$ generate $M$,
there exist $a_{ij}\in A$ ($1\le i\le n$) such that
$y_i=\sum_{j=1}^n a_{ij}x_j$, whence
$$
x_i=\sum_{j=1}^n a_{ij}u(x_j)\qquad\hbox{for $1\le i\le n$}.
\leqno(3.1.1)
$$
Giving $M$ the $A[T]$-module structure defined by $u$,
this implies that
$$
x_i=\sum_{j=1}^n (a_{ij}T)x_j\qquad\hbox{in $M$}.
\leqno(3.1.2)
$$

Now give the $A[T]$-module $M^n$ the Mat$_n(A[T])$-module
structure described in (2.2.4). It is easy to see that
(3.1.2) implies that, in $M^n$,
$$
\pmatrix{
1-Ta_{11} & -a_{12}  & \ldots& -a_{1n}  \cr
-a_{21}   & 1-Ta_{22}& \ldots& -a_{2n}  \cr
\vdots    & \vdots   & \ddots& \vdots   \cr
-a_{n1}   & -a_{n2}  & \ldots& 1-Ta_{nn}\cr
}
\pmatrix{x_1  \cr x_2 \cr \vdots \cr x_n \cr } = 0.
\leqno\rm (3.1.3)
$$

\bye
```

3. Endomorphisms of an *A*-module of finite type

Theorem 3.1. *If M is an A-module of finite type and $u : M \to M$ is an endomorphism of M, the following equivalence holds:*

$$u \text{ is surjective} \iff u \text{ is bijective.}$$

The direction $\Leftarrow$ is obvious. We show the opposite direction $\Rightarrow$. Let $(x_1, x_2, \ldots, x_n)$ be generators for M. Since u is surjective, there exist $y_i \in M$ such that $x_i = u(y_i)$. Since the x_i generate M, there exist $a_{ij} \in A$ $(1 \le i \le n)$ such that $y_i = \sum_{j=1}^{n} a_{ij} x_j$, whence

$$(3.1.1) \qquad x_i = \sum_{j=1}^{n} a_{ij} u(x_j) \qquad \text{for } 1 \le i \le n.$$

Giving M the $A[T]$-module structure defined by u, this implies that

$$(3.1.2) \qquad x_i = \sum_{j=1}^{n} (a_{ij} T) x_j \qquad \text{in } M.$$

Now give the $A[T]$-module M^n the $\text{Mat}_n(A[T])$-module structure described in (2.2.4). It is easy to see that (3.1.2) implies that, in M^n,

$$(3.1.3) \qquad \begin{pmatrix} 1 - Ta_{11} & -a_{12} & \cdots & -a_{1n} \\ -a_{21} & 1 - Ta_{22} & \cdots & -a_{2n} \\ \vdots & \vdots & \ddots & \vdots \\ -a_{n1} & -a_{n2} & \cdots & 1 - Ta_{nn} \end{pmatrix} \begin{pmatrix} x_1 \\ x_2 \\ \vdots \\ x_n \end{pmatrix} = 0.$$

Some variations

After you've run one or both files without errors, try the following variations:

• Replace `\hsize=115mm` at the top of the files with `\hsize=3in`. This resets the width of the page, that is, the length of the lines.

• Add `\vsize=2in` at the top of the files. This sets the height of the page to be only two inches, so you get page breaks.

• Add `\parindent=1in`, or (another time) `\parindent=-1in` at the top. This sets the paragraph indentation.

• Add `\baselineskip=15pt` at the top to set the distance between the bottoms of consecutive lines. (A point, abbreviated pt, is a very commonly used unit in typography; there are about 72 pt in an inch.)

• Add `\parskip=5pt` at the top, to change the spacing between paragraphs.

1.7 Creating a master file

If you have a long document to typeset, it's best not to have all of the text in one file, because it is cumbersome to manipulate big files. A good rule of thumb is that your TₑX files should not exceed 500 lines. So you should have one chapter, or perhaps one section, per file.

Suppose your document is split among four files, called, for example, `doc1.tex`, ..., `doc4.tex`.

It is not necessary to merge the individual files and the macro file to run them together. Instead, it's much better to create a small *master file*, called, for example, `master.tex`. It will say simply this:

```
\input doc1.tex        \input doc3.tex
\input doc2.tex        \input doc4.tex
                       \bye
```

Now when you type `tex master` (or whatever it is that you type to run TₑX on your system), TₑX will read the files `doc1.tex`, ..., `doc4.tex` in sequence, and behave as if the four were one big file.

Suppose also that you want to set several options as you did at the end of the preceding section, for example, the `\hsize`, the `\vsize`, and the `\parindent`. You could start the files `doc1.tex`, ..., `doc4.tex` with the corresponding commands; but since these commands really should affect the whole document, it's best not to encumber the individual files with them—there is no reason they should be read four times. Instead, you can put them all in a *macro file* (or options file, or style file), called, say, `doc.mac`. The file will look something like this:

```
\hsize=4.5in
\vsize=6in
\parindent=1in
```

And now you add `\input doc.mac` at the top of the master file, so it is read just before `doc1.tex`.

1.8 Error messages

It's rare to get a source file entirely right the first time, especially if you are a beginner. When TₑX runs into a place where the file is messed up, it sends you an error message. It also sends the error messages to a `log` file, so you don't have to make a note of them: you can look in the `log` file after the run is over.

Often there isn't much you can do to fix an error at run time, and the best course is to tell TₑX to forge ahead in spite of apparent nonsense. You do this by typing carriage returns as TₑX prints question marks on your screen. On the other hand, sometimes the messages can save a lot of time—as you learn, little by little, what causes them and what they mean.

Here is an error message that you will often encounter in the beginning:

```
! Undefined control sequence
l.27  xxxxxx \hvill
                    yyyyyy
   ?
```

Here xxxxxx and yyyyyy represent the text immediately before and after the error. TEX is saying that it's just read, on line 27, a control sequence \hvill whose meaning it doesn't know. It will ignore it, but it stops to give you a chance to fix things somewhat.

The question mark means that TEX is waiting for a response. Here are five possible responses:

• Typing H or h (followed by CR). TEX offers help, in the form of a somewhat longer explanation of the problem. This option is often useful to beginners.

• Typing CR by itself. This means you have nothing to tell TEX and that it should continue as best it can. Unless the error has subtle ramifications, TEX generally recovers pretty well and subsequent paragraphs are unaffected by the error. But some errors are harder to recover from, and you may find yourself typing CR several times in a row. In this case you can save yourself some trouble by typing Q or q (for "quiet"), which is the same as typing CR to all errors till the end of the run.

• Inserting some text just before yyyyyy. You do this by typing I or i, followed by whatever you want to insert, then CR. In the example above, \hvill was a typo for \hfill, so you can achieve your original intent by typing i\hfill. Watch out: what you type interactively to TEX does not go into your source file! You won't find the corrections there when you edit your text. But the log file does get a copy of everything you type to TEX, so a long definition can be spliced into the source using your editor.

• The next possibility is to type X or x (for "exit"). TEX quits the game, but all pages completed so far are already safely recorded in the dvi file, and can be sent to the printer or the screen.

• In extreme cases you may have to use your operating system's Interrupt or Reset key to abort the run, and try to guess by looking at the source what went so terribly awry. Fortunately this situation is rare. . .

As the \hvill example above shows, an error causes TEX to display the line of the file where it occurs—or rather, a pair of lines, since the line is broken at the point where the error is detected. Errors can be detected not only while new material is being read from a file, but also while TEX is doing something to commands already read (such as "expanding macros"). In this case, several such pairs of lines are shown; only the last one represents what's currently being read from the file. If you get annoyed with unintelligible context lines filling up your screen every time you hit an error, start your file with \errorcontextlines=0. This will inhibit the printing of all context lines but the first (where the proximate cause of the problem lies) and the last (which is the actual line in your file where TEX stumbled). (Note: versions of TEX prior to 3.0 don't know about \errorcontextlines.)

Other common error messages are:

> ! Overfull \hbox   (followed by a dimension)

This is not very serious: the program doesn't even stop running. TEX is about to write on into the margin because it can't fit things into a regular line. If the excess is very big (of the order of 100 pt or more), you may want to investigate if you're using \line or \centerline in the middle of a paragraph. It may also be that TEX got confused some time before and got into math mode by mistake.

! Underfull \hbox   (or Underfull \vbox)

TEX likes neither overfull boxes (the preceding case) nor underfull ones. You can ignore these messages, too: don't worry about the page layout until later.

! Runaway argument?

You gave more than one paragraph to a macro (such as \centerline) that can only handle one paragraph or part of a paragraph at a time.

! Missing $ inserted.

TEX has just entered into math mode on its own, because it's seen something that it should only see in math mode: an exponent, or a Greek letter, for example. So it pretends that the offending object was preceded by a dollar sign. But beware: TEX may know it has to get into math mode, but it won't know when to leave, until it reaches the end of the paragraph! So you'll generally get one or more Overfull \hbox messages, because TEX is typesetting text in math mode, and ignoring spaces as it goes. Don't worry, you'll fix it easily in the source, by just adding dollar signs at the right places.

One hint: when TEX finds an overfull box, it marks it with a thick vertical bar on the right margin. The width of this bar is controlled by \overfullrule , so you can make it invisible by typing \overfullrule=0pt at the top of the file. You can also control how strict TEX is in signaling overfull boxes in the first place: if your text is invading the margin by a negligible amount, say 1 pt, you should set \hfuzz=1pt at the top of the file or, even better, in the macros file, as discussed in section 1.7.

If the file that you're running doesn't end with \end or \bye , TEX will process it and offer you a chance to add more material by prompting you with a * on the screen. If you have nothing to say, type \bye . A CR at this point will make TEX quite frustrated, and generate the message

(Please type a command or say '\end')

If typing \bye or \end won't end the run, it's because the input file left TEX in some strange mode. How to stop in that case depends on your machine: most operating systems have an end-of-file character that you can type to signal that there really is no more input. Consult the documentation that came with TEX, or your local system administrator.

2
The characters of TEX

Here is the list of characters that you can use when typing TEX:

A B ... Z	a b ... z
0 1 ... 9	. , ; : ! ? ' ' "
+ – * / | < > ( ) [ ] @	
$ # % & \ { } ^ _ ~	

These are mostly characters you're accustomed to using: lowercase and uppercase letters, digits, punctuation, accents, and certain mathematical symbols. Of course, you can use the space SP and the carriage return CR.

2.1 Characters that are special to TEX

Look carefully at the last line of the table: it lists the characters that have special meaning to TEX. *These characters don't print as themselves,* but are instead used to communicate with TEX in special ways.

The first four of them sometimes occur in text, and you can get them by preceding them with a backslash: type \$ to get $, \% for %, \& for & and \# for #. The others normally occur only in special situations, for example, braces in mathematics. We will discuss them in later chapters.

The character $ declares that the texts following it should be treated in a special mode, called *math mode.* To get out of this mode, type $ again. It is essential to tell TEX when you're about to type mathematics, because a different set of rules and conventions takes effect then (for instance, spaces are treated quite differently).

Be extra careful with the symbol %! When TEX finds a % not preceded by a backslash, it skips everything remaining on the line, starting with the %. In this way, you can include comments in a TEX file that won't show in the output—a reminder to yourself, an explanation for future reference, or a remark not meant for public eyes. For example, if you type

```
Annual Report              % Judy,
of the Board of Directors  % can you believe
of Graham, Grimm \& Groome % that Steve
% stood me up again last night?
```

TEX "sees" only the following:

```
Annual Report
of the Board of Directors
of Graham, Grimm \& Groome
```

and proceeds to typeset the Annual Report of the Board of Directors of Graham, Grimm & Groome, completely ignoring your justifiable indignation.

Notice the use of \& above in order to get a &. The character &, together with #, is reserved for use in alignments and tables.

Braces, too, deserve special attention. A pair of braces like this:

```
... {...a group...} ...
```

defines a *group*. See chapter 3 for more details.

The characters ^ and _ are used in typing mathematical formulas.

Finally, the character ~ creates a *tie*, or *unbreakable space*. TEX will not break lines at a tie. For example, you should type p.~314 rather than p. 314 in order to prevent an unsightly break. The tie serves another important function: is tells TEX not to leave extra space after the abbreviation, as it generally does following a period. You'll find examples in the first text of section 1.6 and in section 2.5; see also section 5.9.

2.2 Quotes

You should distinguish carefully between opening and closing quotes, ' and '; they are on different keys, ' and '. The position of the closing single quote, or apostrophe, is pretty much standard, but the other depends on what keyboard you're using, and is sometimes hard to find. To get double quotes (of either type) type the corresponding single quotes twice.

So to get

"I've no idea what 'holonomy' means", he said sheepishly.

you should type ``I've no idea what `holonomy' means'', and so on.

The double-quote character " has the same effect as two closing single quotes ''.

2.3 Ligatures and special characters

T_EX treats the sequence of characters ' ' in a special way, and prints a combined character, or ligature, in its place. Here is a complete list of T_EX's ligatures:

ff	⟶ ff	ffi	⟶ ffi	` `	⟶ "	! `	⟶ ¡
fi	⟶ fi	ffl	⟶ ffl	' '	⟶ "	? `	⟶ ¿
fl	⟶ fl	--	⟶ –	---	⟶ —		

There are four types of dashes in T_EX. They are:

- The hyphen, used in compound words, is obtained by typing a single `-`. T_EX inserts a hyphen automatically when it breaks a word between lines.
- The en-dash is a bit longer: –. You get it by typing `--`, and it is used to indicate ranges of numbers; for example, to get pages 13–47 you should type `pages 13--47`.
- The em-dash is even longer—it's used as punctuation, as in this sentence, and you get it by typing `---`.
- The minus sign appears spontaneously in math mode: `$-$` gives $-$.

Whenever you want to typeset plus or minus signs, you should do it in math mode, that is, inside a pair of `$...$`. The result looks much better:

`1+2-3` .... 1+2-3
`$1+2-3$` .... $1 + 2 - 3$

Several other special characters are obtained by typing control sequences:

`\oe , \OE`	⟶ œ, Œ		`\ae , \AE`	⟶ æ, Æ
`\aa , \AA`	⟶ å, Å		`\o , \O`	⟶ ø, Ø
`\l , \L`	⟶ ł, Ł		`\ss`	⟶ ß

To get les œuvres d'Æsope, you should type `les \oe uvres d'\AE sope`. Why are there spaces after the `\oe` and the `\AE`? To tell T_EX where the name of the control sequence ends. *Any number of spaces, and up to one carriage return, are discarded after a control sequence made up of letters.* Here are some more examples:

`Stra\ss burg`	→ Straßburg	`Bergstr\o m`	→ Bergstrøm
`Wroc\l aw`	→ Wrocław	`\AA rhus`	→ Århus

If the space in the middle of a word bothers you, you can delimit the control sequence in another way: `Stra{\ss}burg`, `Bergstr{\o}m`, and so on. Here the `}` also tells T_EX that the control sequence has ended.

Suppose you're a physicist, and want to typeset

Light with a wavelength of 3000 Å is invisible.

If you type `...\AA is invisible`, you get ... Åis invisible: not at all what you want! For the space not to be discarded, you must make sure it doesn't come right after the control sequence name. You can choose according to taste:

`{\AA}` is invisible   or   `\AA\` is invisible.

2.4 Accents

To get an accent above a letter, type the appropriate control sequence *before* the letter: for example, é is obtained by typing `\'e`. Any letter or symbol can follow an accent. Here is a list of the available accents, and how they look on the letter 'o':

`\`o`	ò	grave accent		`\'o`	ó	acute accent
`\^o`	ô	circumflex accent		`\"o`	ö	dieresis or umlaut
`\~o`	õ	tilde		`\u o`	ŏ	breve
`\=o`	ō	macron or bar		`\.o`	ȯ	dot accent
`\b o`	o̲	bar-under accent		`\d o`	ọ	dot-under accent
`\t o`	o͡o	tie-after accent		`\v o`	ǒ	háček or check
`\H o`	ő	Hungarian umlaut		`\c o`	o̦	cedilla

Notice the difference in syntax between `\`o` and `\.o`, on the one hand (no space after the accent), and `\b o`, `\v o` on the other (space required). Here are some examples:

`l\`ese-majest\'e`	→ lèse-majesté	`Fr\"aulein`	→ Fräulein
`\v Cekoslovensko`	→ Čekoslovensko	`na\"\i ve`	→ naïve
`Tr\^es Cora\c c\~oes`	→ Três Corações	`na\"{\i}ve`	→ naïve

When an 'i' gets an accent, it should first be deprived of its dot. That's why we use the control sequence `\i` in the last two lines above. As usual, we have to leave a space after it or enclose it in braces. The same remarks apply to 'j', whose dotless version is obtained by typing `\j`.

If you need to use certain accents a lot—if you're writing a text in French, say—it's possible to arrange things so that fewer keystrokes are needed for each accent. For instance, you can *redefine* `"` so that it stands for the dieresis, rather than for quotation marks. You can even make TEX look ahead and replace an accented 'i' by 'ï', so you would just type `na"ive` to get naïve. We'll come back to this point in section 12.6.

2.5 Two exercises

Test yourself by trying to typeset this bibliographical reference in Rumanian, without looking at the solution below:

GEORGESCU, V. A., Bizanţul şi instituţiile româneşti pînă la mijlocul secolului al XVIII-lea (Byzanz und die rumänishen Institutionen bis zur Mitte des 18 Jahrhunderts). Bucureşti, Ed. Academiei RSR, 1980. Reviewed by C. R. Zach, in Südost-Forschungen, 40, 1981, pp. 434–435.

```
GEORGESCU, V. A., Bizan\c tul \c si institu\c tiile
rom\^ane\c sti p\^\i n\v a la mijlocul
secolului al XVIII-lea (Byzanz und die rum\"anishen
Institutionen bis zur Mitte des 18 Jahrhunderts).
Bucure\c sti, Ed.~Academiei RSR, 1980.
Reviewed by C.~R.~Zach, in S\"udost-Forschungen,
40, 1981, pp.~434--435.
```

And here is a short text in Berber, a North African language. To get γ and ε you can use the abbreviations \g and \e, after having defined them in the following way: \def\e{ε} \def\g{γ}.

Yenna-yas: Ṛuḥ ẓẓu lejnan. Ma d ayefki, yenna-yas: ṛuḥ aγtixsi tamezgult, zlu-ţ, tazuḍ-ţ, twenεeḍ-ţ irkweli, tawiḍ-ţ γer uqemmuc l-lγaṛ t-tizemt.

```
Yenna-yas : \d Ru\d h \d z\d zu lejnan. Ma d ayefki,
yenna-yas: \d ru\d h a\g tixsi tamezgult, zlu-\c t,
tazu\d d-\c t, twen\e e\d d-\c t irkweli, tawi\d d-\c t
\g er uqemmuc l-l\g a\d r t-tizemt.
```

3
Groups and modes

3.1 Groups

What is a group?

You create a group when you put text within braces:

```
... {AAA...{BBB...bbb}} ... {CCC...ccc} ...
```

Remember that the braces don't show on the output. They simply mark the beginning and the end of the group. In the example above there are three groups. The first starts with `{AAA...`, the second is `{BBB...bbb}`, and the third `{CCC...ccc}`. The group `{AAA...}` contains the group `{BBB...bbb}`.

You also create a group whenever you use one of the box commands `\hbox{...}`, `\vbox{...}`, `\vtop{...}` and `\vcenter{...}`, as well as many others like `\halign{...}` and `\matrix{...}`. The general rule is that whenever you see braces, there is a group there. Pairs of single dollar signs `$...$` and double dollar signs `$$...$$`, used to delimit math mode, also define groups.

Groups must be correctly nested: each open brace must be balanced by a close brace, and a group cannot encroach on another. It two groups overlap, one must be entirely inside the other. In this example:

```
{AAA...{BBB...{CCC...bbb}...ccc}...aaa}
```

we've tried to interweave the groups `{BBB...bbb}` and `{CCC...ccc}`, but the result is a group `{AAA...aaa}` containing a group `{BBB...ccc}` containing a group `{CCC...bbb}`. And this example

```
{AAA...{BBB...$CCC...bbb}...ccc$...aaa}
```

gives an error message, because we're trying to close the group {BBB...bbb} before closing the group $CCC...ccc$ that began inside it.

One can also define groups using certain control sequences:

\bgroup...\egroup   or   \begingroup...\endgroup

The usefulness of these constructions will become clear later on.

What are groups for?

TEX keeps track of a great number of variables: fonts, margins, interline spacing, and many others. Consider, for example, the variable \parindent , which says by how much the first line of each paragraph should be indented. (It is also used by such macros as \item and \narrower.) The value of this variable when you start plain TEX is 20 pt, or about .27 inches; suppose you want to temporarily make it zero so as to get a few unindented paragraphs. You can write \parindent=0pt and, after you're done, cancel the change by writing \parindent=20pt . But this is inconvenient for two reasons: you must know the original value of the variable, and you mustn't forget to change it back. A more reliable method is to use a group:

{\parindent=0pt ...}

Inside the group, the \parindent is 0 pt. After the group is finished, the value of \parindent is automatically reset to what it was before; in general, *values assigned to variables within a group are automatically forgotten when TEX steps out of the group.* Such variables then revert to the values they had before the group was entered; we say that their values were *local* to the group. Variables that don't receive a new value inside the group, of course, have the same value throughout. (Those familiar with the computer science jargon will realize that grouping is handled by a stack mechanism.)

This is an important point, and it affects not only variables but all aspects of TEX: macros, fonts, etc. Here is a little example to make it sink in. The big rectangle represents the page, and the small rectangle the group:

```
\def\toto{Hello, world!}
\parindent=20pt
\parskip=0pt

...

   {\font\eightrm=cmr8
   \def\toto{Good bye!}
   \parindent=0pt
   ...}

...
```

Before and after the group, the macro \toto stands for Hello, world! , the \parindent is 20 pt and the \parskip is 0 pt. Inside the group, \toto stands for Good bye!, the \parindent is 0 pt and the \parskip is still 0 pt. The font command \eightrm (explained in the next chapter) works inside the group, but is undefined outside.

As usual, there is a way to tell TEX to act differently. To make \parindent retain its value after the end of the group, you can say {\global\parindent =2cm...}. Similarly, if you want the new definition of \toto to remain after the group, entirely overriding the old one, you can say

$$\text{\{\textbackslash global\textbackslash def\textbackslash toto\{Good bye!\}...\}.}$$

Now, as promised, a few words about the commands \bgroup ... \egroup and \begingroup ... \endgroup. The pair \bgroup and \egroup are just new names for braces. This means you can replace an open brace by a \bgroup and a close brace by an \egroup, if you feel like it. This in itself is not very useful, until you find yourself trying to define a macro that has unbalanced braces inside. (See section 12.2 for an example.)

The other pair, \begingroup and \endgroup, is not synonymous with braces: a group starting with \begingroup must end with \endgroup. Why this distinction? Because it allows one to build mechanisms to check for errors much more thoroughly than can be done in plain TEX. (Needless to say, creating such mechanisms is not a task for beginners.) For example, LATEX uses \begingroup and \endgroup to define *environments* that begin and end with paired commands. If your file has something like \begin{document} ... \begin{theorem} ... \end{document}, you probably left something out inadvertently, and an error message is generated saying that the \begin{theorem} doesn't have a matching \end{theorem}. In plain TEX this sort of error is much harder to detect.

3.2 Modes

The three modes of TEX

In chapter 8, you will see that TEX is mostly concerned with boxes. Roughly speaking TEX places these boxes on the page in one of three ways:

• side by side, like a typesetter setting a literary text or someone stringing beads (*horizontal mode*);

• on top of one another, like a storekeeper stacking up boxes of merchandise (*vertical mode*);

• according to special rules, like a typesetter setting a mathematics text (*math mode*).

When TEX is in horizontal mode, it places characters, or boxes, side by side. Horizontal mode has two submodes:

• *ordinary horizontal mode*. TEX is in this mode when it is setting a paragraph. It puts all the characters and boxes side-by-side into a long horizontal list. Later, when it hits the end of the paragraph, it breaks up the list into *lines* and stacks them up.

- *restricted horizontal mode.* TEX is in this mode when it's setting text inside an \hbox or \halign. The behavior is almost the same as in ordinary horizontal mode, but the horizontal list that TEX creates is not broken into lines—it turns into a box as wide as the sum of its components.

When TEX is in vertical mode, it stacks boxes on top of one another. Vertical mode, too, has two submodes:

- *ordinary vertical mode.* TEX is stacking up boxes (for example, lines of text) at the "outer level," to build up a page. When the boxes pile up to the height of a page, TEX ships them out and begins a new page.

- *internal vertical mode.* TEX is stacking up boxes to make another box, rather than a page. The difference here, like the difference between the two horizontal modes, is that in internal vertical mode will keep stacking boxes up to matter how tall the pile gets.

Math mode, too, has (you guessed it) two submodes:

- *text math mode*, which begins and ends with single dollar signs $...$. In this mode TEX builds up a math formula to be included inside a line of text.

- *display math mode*, delimited by double dollar signs $$...$$. In this mode TEX builds up a math formula to be displayed, that is, centered on a line by itself.

Changing modes

When it starts, TEX is in ordinary vertical mode. Any character, or any of several commands that indicate that a paragraph is about to start, makes TEX go into horizontal mode. Dollar signs make it go into math mode. The end of a paragraph, signaled by a blank line or by \par, makes it go back to vertical mode, as does any intrinsically vertical command.

Why talk about modes?

Normally, you won't have to worry about what mode TEX is in. But the same character or command can cause different reactions when encountered in different modes, and this sometimes causes unexpected behavior or error messages. For example, if you type {\obeyspaces   a} in the middle of a paragraph, you get several spaces before the 'a', but if you do it right after a blank line you get none! The reason is that after a blank line you're in vertical mode, a paragraph having just ended; spaces have no effect in vertical mode or math mode.

Some commands simply don't make sense in some modes. In restricted horizontal mode TEX is meant to create a single line, so naturally it doesn't expect to encounter \vskip, a command to add vertical spacing, and complains if it does. For a similar reason, TEX ignores (although it doesn't consider it an error) a blank line inside an \hbox.

Here is a practical example: as it finishes processing your file, TEX reads the final \bye command and produces the message:

```
! You can't use '\end' in internal vertical mode.
```

What's happening? When it read `\bye` (which includes `\end` in its definition), TEX was in internal vertical mode—inside a `\vbox` or some such thing. It hadn't finished the box yet, probably because a right brace was missing somewhere. So the solution is to add the right brace and repeat the `\bye` command, by typing

`i}\bye` .

3.3 For the aspiring wizard

If you want to see TEX change modes, start TEX interactively (without a file name) and type the lines below one by one (left column, then right column). You'll see a message on the screen every time the mode changes.

```
\tracingcommands=1        \vbox{
\tracingonline=1          b\par
\hbox{                    c}
a                         }
$x$                       \bye
```

4
The fonts TeX uses

4.1 TeX's fonts

Plain TeX allows you to use several types of text fonts: roman, italic, boldface, slanted and typewriter. (Slanted and italic are altogether different things: you'll see samples soon.) All these fonts belong to a big family, called Computer Modern, especially designed by Knuth for use with TeX. Most systems that run TeX offer all the Computer Modern fonts, which in addition to the styles above include more esoteric ones such as sans-serif, unextended bold and unslanted italic.

Nowadays many installations of TeX, both on PC's and on bigger systems, allow access to PostScript fonts: Times, Helvetica, and whatever else your printer has. You can also get fonts for foreign alphabets, special characters, and so on.

The fonts that you can use in TeX fall into two categories: those already known when TeX starts—they're called preloaded—and those that you have to tell TeX about. Let's look at them in turn.

4.2 Preloaded fonts

When you start plain TeX, unless you say something to the contrary, you get Computer Modern Roman at 10 point. This paragraph is set using this font: it looks somewhat like the Times Roman font used for the rest of this book, but the face is more open, more rounded and less tall.

TeX knows this font under the name `\tenrm`, and the same roman font in 7 and 5 point size is known as `\sevenrm` and `\fiverm`. That's pretty self-explanatory!

To use 7-point roman, it's enough to type `\sevenrm` : from then on, that's the font you get. Let's have a bit of fun:

He had a great big head
 on an average body,
 and short, spindly legs.
	`He had a great big head\par`
	 `\sevenrm on an average body,\par`
	 `\fiverm and short, spindly legs.`

(Notice the use of `\par` to start a new line.)

Now say you have just a bit of text to set in 7 point:

She tiptoed quietly into the room.

You could type `She tiptoed \sevenrm quietly \tenrm into the room.` But it's annoying to have to revert to `\tenrm` explicitly. There is a better alternative: place the 7-point word and the font-change command inside a group, that is, within braces. When TEX leaves the group, it restores automatically the previous font.

`She tiptoed {\sevenrm quietly} into the room.`

Here is the complete list of non-math fonts preloaded in plain TEX; see the next section for an explanation of the file name column.

font file name	font-change command	description
cmr10	\tenrm	10-point roman
cmr7	\sevenrm	7-point roman
cmr5	\fiverm	5-point roman
cmbx10	\tenbf	10-point bold
cmbx7	\sevenbf	7-point bold
cmbx5	\fivebf	5-point bold
cmti10	\tenit	10-point italic
cmsl10	\tensl	10-point slanted
cmtt10	\tentt	10-point typewriter

In addition to the font-change commands above, plain TEX offers abbreviations `\rm`, `\bf`, `\it`, `\sl` and `\tt` which give the corresponding 10-point fonts. They also work correctly in math mode, which is not the case for `\tenrm`, etc. (see section 4.7 if you must know why). These abbreviations, then, are the preferred way to change fonts, unless you need a size change.

4.3 Loading other fonts

Suppose you want to use 8-point roman, a font that is not preloaded in plain TEX. You must define a name to refer to that font—"register" it with TEX, so to speak— and also tell TEX where to find information about it. Information about each font is kept in a special file somewhere on your computer system; for Computer Modern Roman at 8 point, this file is called `cmr8`. The registration command in this case is

<code>\font\eightrm=cmr8</code>

(No backslash before the file name!) Once the registration formalities are over, TEX will treat the newly defined `\eightrm` just like the predefined names of the previous section:

He had a great big head	`He had a great big head\par`
on an average body,	`\eightrm on an average body,\par`
and short, spindly legs.	`\sevenrm and short, spindly legs.`

There is nothing sacred about the name `\eightrm`: you could have called this font `\romaneight`, or `\romanVIII` (sorry, no digits allowed), or `\romainhuit` if you're more fluent in French. The notary that works inside TEX will register the most outlandish names without batting an eyelash.

A common mistake among beginners: registering a new font, that is, saying `\font \toto=`..., doesn't mean that TEX switches to that font right away. It just learns its name. You must still type `\toto` to switch to the new font.

PostScript fonts

Assuming that your installation of TEX supports PostScript fonts, you must load them into TEX just the same as any other font that is not preloaded. The only thing to watch out for is that, unlike Computer Modern fonts, PostScript fonts may have different names and design sizes on different systems. To get Times Roman at 10 pt, the main text font in this book, you might have to say

<code>\font\timesX=PS-Times-Roman</code>

on one system, while on another the right incantation might be

<code>\font\timesX=Times at 10pt</code>

because the "design size" is arbitrarily set at 1 pt. You shouldn't use `\times` as the name of a font, because this control sequence already means the symbol ×.

4.4 A cornucopia of fonts

Here are some samples of Computer Modern fonts. First the roman (`cmr`) with two simple variations: boldface (`cmbx`, for *bold extended*, a reference to its width) and caps–small caps (`cmcsc`):

`cmr10`   MURPHY'S LAW: If anything can go wrong, it will.

`cmbx10`   **MURPHY'S LAW: If anything can go wrong, it will.**

`cmcsc10` MURPHY'S LAW: IF ANYTHING CAN GO WRONG, IT WILL.

Then come the italic fonts, available in regular (`cmti`, for *text italic*) and bold (`cmbxti`):

`cmti10`   *MURPHY'S LAW: If anything can go wrong, it will.*

`cmbxti10`   ***MURPHY'S LAW: If anything can go wrong, it will.***

Tilting an upright font, without any other modifications, gives you a slanted one. Notice how slanted differs from italic.

cmsl10   *MURPHY'S LAW: If anything can go wrong, it will.*

cmbxsl10   ***MURPHY'S LAW: If anything can go wrong, it will.***

The basic typewriter font (cmtt) has several variations: italic (cmitt), slanted (cmsltt) and caps–small caps (cmtcsc):

cmtt10   `MURPHY'S LAW: If anything can go wrong, it will.`

cmitt10   *`MURPHY'S LAW: If anything can go wrong, it will.`*

cmsltt10   *`MURPHY'S LAW: If anything can go wrong, it will.`*

cmtcsc10   `MURPHY'S LAW: IF ANYTHING CAN GO WRONG, IT WILL.`

Here are several sans-serif fonts (dc stands for *demibold condensed*):

cmss10   MURPHY'S LAW: If anything can go wrong, it will.

cmssbx10   **MURPHY'S LAW: If anything can go wrong, it will.**

cmssdc10   **MURPHY'S LAW: If anything can go wrong, it will.**

cmssi10   *MURPHY'S LAW: If anything can go wrong, it will.*

Most of the fonts above are available in a range of sizes, although we chose to stick to 10 point for these samples. There are still other Computer Modern fonts, like cmdunh (for Dunhill) and cmff (for... funny), as well as fonts with mathematical symbols:

cmdunh10   MURPHY'S LAW: If anything can go wrong, it will.

cmff10   MURPHY'S LAW: If anything can go wrong, it will.

Stepping out of Computer Modern we have Euler Fraktur, distributed by the American Mathematical Society

eufm10   𝔐𝔘ℜ𝔓𝔥𝔜'𝔖 𝔏𝔄𝔚: 𝔍𝔣 an𝔫thing can go 𝔴𝔯ong, it 𝔴ill.

and many fonts for foreign alphabets: Cyrillic, Greek, Hebrew... We've barely scratched the surface.

4.5 Scaling of fonts

Assume you need a 12-point roman font, but your TₑX installation only has the 10-point font cmr10 . You can scale up the existing font by saying

> `\font\twelverm=cmr10 at 12pt`

This tells TₑX to multiply all the dimensions of font cmr10 by a factor of $12/10 = 1.2$, resulting, in effect, in a 12-point font. (TₑX knows that the original size, or *design size*, of cmr10 is 10 pt because this information is written in the tfm file; the name of the file is irrelevant.)

The following construction gives exactly the same result:

> `\font\twelverm=cmr10 scaled 1200`

(notice there is no backslash before `scaled` or `at`). The keyword `scaled` works like this: `scaled 1000` means a scaling factor of 1 (that is, no change), `scaled 1200` a factor of 1.2 (that is, a 20% increase in size), `scaled 500` a factor of .5 (that is, a reduction by 50%), and so on.

To avoid useless calculations, plain T_EX has several control sequences that you can use after `scaled`: `\magstep1` stands for $1000 \times 1.2 = 1200$, `\magstep2` for $1000 \times 1.2 \times 1.2 = 1440$, and so on up to `\magstep5`. Using these conventions, the definition above could also be written

$$\texttt{\textbackslash font\textbackslash twelverm=cmr10 scaled \textbackslash magstep1}$$

There is also `\magstephalf`, which means a magnification factor of $\sqrt{1.2} = 1.09545$.

You can blow fonts up or down at will as far as T_EX is concerned, since T_EX doesn't know or care anything about fonts except for their dimensions. But you'll run into trouble if the driver program that send the characters to the printer can't find information about them at the requested magnifications. In other words, if your system doesn't have a file containing the bitmaps for `cmr10` at 12 point, which are different from those at 10 point, T_EX won't complain about the definition of `\twelverm`, but the printer driver will either ignore all characters from that font, or try to replace them by some approximation based on the `cmr10` bitmaps that it does have. The result varies from printer to printer, but it's always less than ideal. This is not a problem for PostScript fonts, which are outline fonts, and can be magnified or reduced without worries about the driver. Here are some more examples of PostScript font registering:

$$\texttt{\textbackslash font\textbackslash tenpalatino=Palatino at 10pt}$$
$$\texttt{\textbackslash font\textbackslash bighelvetica=Helvetica at 30pt}$$

One last thing: each scaled version of a font must be registered separately. For example, saying `\font\twelvebf=cmr10 at 12pt` does not entitle you to use `cmr10` at the design size of 10 pt. In other words, a font registration command associates with a control sequence a pair (file name, scaling factor), and that pair only.

4.6 Global scaling

The next command isn't, properly speaking, a font command: it allows you to magnify or shrink a whole document. By typing

$$\texttt{\textbackslash magnification=1200}$$

at the beginning of your document, you'll magnify it by 20%. There is an important restriction: you can only set the global magnification once, at the beginning of the run. It is illegal to change it along the way, though you can still scale individual fonts.

The effects of `\magnification` are combined with those of the individual scaled font definitions. For example, if you use the `\twelverm` font defined above with

\magstep1 in a document that starts with \magnification=\magstep1 , the result will be a font scaled up by a factor of 1.2×1.2, or 44%. This, by the way, is why the \magstep series is multiplicative: the cumulative effect of two \magstep1 is a \magstep2 , and so on.

The \magnification command is very useful in proofreading a manuscript. For example, this book is set in 10 point, but all proofs were printed bigger by the use of the command \magnification=\magstep1 . At the typesetter, this command was removed and the final copy came out as you see it.

The truth of the matter

Suppose you want to set your document in 12 point. Many systems don't have fonts like cmr12 , etc., because big fonts take up a lot of space; but they probably do have cmr10 at 12 point. So you have two choices: you can either scale each font individually, or, much more conveniently, start your document with the command \magnification=\magstep1 .

The problem then is that all dimensions are increased by 20%. If you want to leave 1 inch between lines, you have to divide 1 inch by 1.2 and type \vskip .833in . Not at all fun!

Fortunately, there is a better solution:

$$\hskip 1truein$$

The prefix true can be written before any of TEX's units (in , cm , pt , pc , etc.; more about them later). When TEX encounters a true , it divides the dimension by the current global magnification before using it. It performs the calculation you'd have to do otherwise.

4.7 For the aspiring wizard

Fonts in math mode

TEX has a sophisticated mechanism for handling fonts in math mode. It automatically chooses a smaller size for a character that is subscripted or superscripted to another character, and an even smaller size for the subscript or superscript of another subscript or superscript:

$2A^2$, nA_n .... $2A^2, nA_n$

$R+B^{2^S}$, $S-B^{R_S}$ .... $R + B^{2^S}, S - B^{R_S}$

$R+B_{2^S}$, $S-B_{R_S}$ .... $R + B_{2^S}, S - B_{R_S}$

The upshot of this intricate mechanism is that, while in normal text there is the notion of a single current font, in math mode we have instead a current family of similar-looking fonts in three different sizes, for example, 10, 7 and 5 point. A family of math fonts is referred to by the control sequence \fam , followed by a number, like \fam0 . The three members of \fam0 are called \textfont0 , \scriptfont0 and \scriptscriptfont0 .

On page 351 of *The TEXbook* you will find the following code:

```
\textfont0=\tenrm \scriptfont0=\sevenrm
\scriptscriptfont0=\fiverm
\def\rm{\fam0\tenrm}
```

The commands on the first two lines populate the family `\fam0`. The last line says that, when TEX encounters `\rm`, the current text font becomes `\tenrm` and the current math family becomes `\fam0`. Outside math mode, then, TEX will use the font `\tenrm`, while inside math mode it will use the fonts of `\fam0`: namely, `\tenrm` for "normal" stuff, `\sevenrm` for subscripts and superscripts, and `\fiverm` for second-order subscripts and superscripts:

`$\rm S-B_{R_S}$` . . . . $S - B_{R_S}$

It is important to understand that a name registered with the `\font` construction has no effect whatsoever within math mode. The only way to change fonts in math mode is to go through a `\fam` construction, or using an `\hbox`, which temporarily puts you in horizontal mode:

`$A$, $\eightrm A$, {\eightrm $A$, $\hbox{A}$}` . . . . A, A, A, A

Back to plain TEX's definitions. Here is the family `\fam1`, which describes the special italic fonts used in mathematical formulas:

```
\textfont1=\teni \scriptfont1=\seveni
\scriptscriptfont1=\fivei
\def\mit{\fam1}  \def\oldstyle{\fam1\teni}
```

The command `\mit` does nothing outside math mode, because it changes the math family but not the current text font. On the other hand, `\oldstyle` works both inside and outside math mode. The reason for two commands is that math italic letters are not meant to be used outside math mode, but the digits in the same font, 0123456789, can be so used. So a user typing `\oldstyle` need not be aware that TEX is switching to the math italic font.

Families `\fam2` and `\fam3` describe the mathematical symbols and extensible symbols that TEX uses. We won't go into the details of them.

Let's turn now to the definition of the `\it` macro:

```
\newfam\itfam \textfont\itfam=tenit
\def\it{\fam\itfam\tenit}
```

Families 0 through 3 have a special meaning to TEX and are generally referred to by number. But after that, remembering the numbers of families becomes a chore—what was that `\fam4` again?—so TEX provides a symbolic way to refer to them. The command `\newfam\itfam` announces that from now on there is a new family, whose number is `\itfam`. Behind the scenes, TEX assigns the value 4 to `\itfam`, but we don't have to worry about that—we just type `\fam\itfam`, which is much more expressive than `\fam4`. (Nor is the `fam` in `\itfam` obligatory; `\newfam\toto` and `\fam\toto` would do just as well.)

The definition of the \it macro is such that $\it A$ works (the 'A' is set in 10-point text italic), but $\it A_k$ doesn't, because the subscript 'k' has no associated font. It causes the error message

> ! \scriptfont 4 is undefined (character k).

(Unfortunately here TEX does not use the symbolic name for the family...)

To make sure you got everything, let's look at the definition of the \bf macro:

```
\newfam\bffam \textfont\bffam=\tenbf
\scriptfont\bffam=\sevenbf
\scriptscriptfont\bffam=\fivebf
\def\bf{\fam\bffam\tenbf}
```

Here all three members of the family are defined: you can use \bf anywhere in a mathematical formula.

Defining new font families

We now know enough to create our own font-change macros, parallel to \rm, \bf, and so on. For example, assuming that PostScript Times Roman is available in the file Times (cf. the end of section 4.3), we can create a macro \tm that switches to Times in math mode as well as in text:

```
\font\tentm=Times at 10pt \font\seventm=Times at 7pt
\font\fivetm=Times at 5pt \newfam\tmfam
\textfont\tmfam=\tentm    \scriptfont\tmfam=\seventm
\scriptscriptfont\tmfam=\fivetm
\def\tm{\fam\tmfam\tentm}
```

The American Mathematical Society, or AMS, distributes a set of fonts containing, among other things, Fraktur or "gothic" fonts (eufm) and the "blackboard bold" some mathematicians are fond of: $\mathbb{A}, \mathbb{B}, \ldots, \mathbb{Z}$ (msbm fonts; see the Dictionary under \bb). Here is, for the sake of completeness, the definition of a macro \frak that switches to Fraktur fonts both in math mode and in text:

```
\font\tenfrak=eufm10      \font\sevenfrak=eufm7
\font\fivefrak=eufm5      \newfam\frakfam
\textfont\frakfam=\tenfrak \scriptfont\frakfam=\sevenfrak
\scriptscriptfont\frakfam=\fivefrak
\def\frak{\fam\frakfam\tenfrak}
```

If you want to limit Fraktur fonts to math use, the last line should read

```
\def\frak{\fam\frakfam}
```

Size-change commands

Plain TEX basically works with only one font size: 10 point. The 7-point and 5-point fonts it defines are for use in subscripts and superscripts of 10-point math formulas.

One of the commonest needs of even a novice user of TEX is for a command that changes the size of all fonts in a coherent manner—for instance, to make footnotes smaller than the text. Such a command should set things up so that, conceptually, \bf , \it , etc., as well as all math constructions, work exactly as before, but the fonts used are appropriately smaller.

We start by collecting together all of plain TEX's definitions that have to do with font changes, and put them into a macro that we call \tenpoint . (Actually, we'd want \tenpoint to take care of other things as well, like interline spacing. We won't go into this now; see the Dictionary for a complete listing.)

```
\def\tenpoint{%
\textfont0=\tenrm \scriptfont0=\sevenrm
\scriptscriptfont0=\fiverm \def\rm{\fam0\tenrm}%
\textfont1=\teni \scriptfont1=\seveni
\scriptscriptfont1=\fivei \def\oldstyle{\fam1\teni}%
\textfont2=\tensy \scriptfont2=\sevensy
\scriptscriptfont2=\fivesy
\textfont\itfam=\tenit \def\it{\fam\itfam\tenit}%
\textfont\slfam=\tensl \def\sl{\fam\slfam\tensl}%
\textfont\ttfam=\tentt \def\tt{\fam\ttfam\tentt}%
\textfont\bffam=\tenbf \scriptfont\bffam=\sevenbf
\scriptscriptfont\bffam=\fivebf  \def\bf{\fam\bffam\tenbf}%
\rm}
```

There are several things to observe here:

• The % at the end of certain lines is necessary in order to prevent the CR from creeping into the definition; otherwise they would appear as spurious spaces when \tenpoint is called in horizontal mode. Only following a control sequence made up of letters are CR and SP harmless.

• The members of family \fam3 are not redefined: extensible symbols figure out their own size from the context.

• Font names like \tenrm and family names like \itfam are defined by plain TEX once and for all, and should not be redefined inside \tenpoint .

• The last line of the definition is a call to \rm , so when you type \tenpoint , you get 10-point roman by default. Naturally, you can choose a different default by changing this line.

• You can include in the definition of \tenpoint your own font-change commands such as \tm and \frak . Again, commands such as \newfam\tmfam and \font\tentm=Times at 10 pt should not go inside \tenpoint , but before, so they're seen only once.

We're now ready to switch over to 8 point, by defining a macro \eightpoint in every way analogous to \tenpoint :

```
\def\eightpoint{%
\textfont0=\eightrm \scriptfont0=\sixrm
\scriptscriptfont0=\fiverm \def\rm{\fam0\eightrm}%
\textfont1=\eighti \scriptfont1=\sixi
```

```
\scriptscriptfont1=\fivei \def\oldstyle{\fam1\eighti}%
\textfont2=\eightsy \scriptfont2=\sixsy
\scriptscriptfont2=\fivesy
\textfont\itfam=\eightit \def\it{\fam\itfam\eightit}%
\textfont\slfam=\eightsl \def\sl{\fam\slfam\eightsl}%
\textfont\ttfam=\eighttt \def\tt{\fam\ttfam\eighttt}%
\textfont\bffam=\eightbf \scriptfont\bffam=\sixbf
\scriptscriptfont\bffam=\fivebf \def\bf{\fam\bffam\eightbf}%
\rm}
```

We also need, somewhere outside the definition of \eightpoint, some incantations to help TEX to place accents correctly in math mode:

```
\skewchar\eighti='177 \skewchar\sixi='177
\skewchar\eightsy='60 \skewchar\sixsy='60
```

Let's see how we dethroned the omnipresent Computer Modern text fonts in favor of PostScript Times in this book. All we had to do was redefine the change commands \it and \bf to call their corresponding Times counterparts, and define a new command \tm, as explained above, that calls Times roman.

```
\font\tentm=Times at 10pt     \def\tm{\fam0\tentm}
\font\tentmit=TimesI at 10pt \def\it{\fam\itfam\tentmit}
\font\tentmbf=TimesB at 10pt \def\bf{\fam\bffam\tentmbf}
```

We kept the old \rm, to make it easier to give examples of TEX output. We also kept the Computer Modern math fonts, by not redefining the families \fam0, \itfam and \bffam. The reason is that PostScript fonts do a poor job in math mode: the spacing is wrong and some characters are simply not available.

Naming a character

Page 427 of *The TEXbook* shows a table of the 128 characters of font cmr10. The position of character ß is 25, also expressed as octal 31 ($3 \times 8 + 1 = 25$) or hexadecimal 19 ($1 \times 16 + 9 = 25$). The low-level command to access this character is {\tenrm\char25} or {\tenrm\char'31} or {\tenrm\char"19}, depending on what base you prefer to work with. (Note that the hex number is preceded by a double quote " , not two single quotes '' .)

A higher-level command to print the character in position 25 of the current font could be defined as

\def\ss{\char25} or \def\ss{\char'31} or \def\ss{\char"19}.

But there is an alternative, more efficient, command:

```
\chardef\ss=25  or  \chardef\ss='31  or  \chardef\ss="19 .
```

The analogous commands for characters to be used in math mode are \mathchar and \mathchardef. Their use is somewhat complicated by the need to specify what family a character belongs to and what purpose it will serve (i.e., whether it's an ordinary character or an operator or punctuation), because the amount of space placed around it depends on this. For details, see section 11.2 and the Dictionary.

4.8 Exercise

A judicious choice of fonts can make bibliographical references such as the one in section 2.5 much more readable. Using the font-change commands of this chapter, format that Rumanian reference according to one style often used in bibliographies:

GEORGESCU, V. A., *Bizanţul şi instituţiile româneşti pînă la mijlocul secolului al XVIII-lea* (Byzanz und die rumänishen Institutionen bis zur Mitte des 18 Jahrhunderts). Bucureşti, Ed. Academiei RSR, 1980. Reviewed by C. R. Zach, in *Südost-Forschungen*, **40**, 1981, pp. 434–435.

Here's one possible solution:

```
\font\sc=cmcsc10
{\sc Georgescu, V. A.}, {\it Bizan\c tul \c si institu\c tiile
rom\^ane\c sti p\^\i n\v a la mijlocul
secolului al XVIII-lea\/} (Byzanz und die rum\"anishen
Institutionen bis zur Mitte des 18 Jahrhunderts).
Bucure\c sti, Ed.~Academiei RSR, 1980.
Reviewed by C. R. Zach, in {\it S\"udost-Forschungen},
{\bf 40}, 1981, pp.~434--435.
```

One control sequence in this code hasn't been discussed before: \/. If you have an italicized *word* before one in an upright font, the two often appear too close together, because of the first word's slant. The previous sentence, for instance, was typed with ... {\it word} before ... If we had said instead

... {\it word\/} before ...

the result would have been better: *word* before. Generally, then, this italic correction should be used whenever there's a switch from a slanted to an upright font.

5
Spacing, glue and springs

What sets apart a truly beautiful typesetting job is the treatment of white space! For this reason TeX has a rich set of commands devoted to the control of spacing.

5.1 Horizontal spacing

The space bar and the carriage return key

We saw in section 1.6 that TeX gives special treatment to spaces and carriage returns (represented by SP and CR) in the sense that they don't always appear in the printed output. Here are the rules again:

- Several consecutive spaces in the input file produce only one space in the printed document.

- A single carriage return is equivalent to a space and produces one space in the printed document—in particular, it absorbs spaces at the end of the preceding line and at the beginning of the following one.

- Two or more carriage returns in a row, that is, one or more blank lines, start a new paragraph.

- One or more spaces or a single carriage return after a control sequence made up of letters don't produce any spaces in the output. They merely indicate the end of the control sequence name. For example: `\OE dipus` and `\TeX book` give Œdipus and TeXbook.

This last rule was discussed briefly in section 2.3, and we saw there that to print TeX makes nice formulas we must type

```
{\TeX} makes nice formulas   or   \TeX\ makes nice formulas .
```

The `\ ` control sequence—a backslash followed by one or more spaces—forces TeX to produce a space; it works both in horizontal mode and in math mode. So in order to have two spaces between XXX and YYY, you can type `XXX\ \ YYY` or `XXX \ YYY`. But this isn't really the best way to do it; the command `\hskip`, explained below, is preferable.

Unbreakable spaces

As we saw in section 2.1, the tilde ˜ has a special meaning to TEX: it represents a tie, that is, a space where no line break is allowed. For example, you should type `D.˜Knuth`, and `pp.˜10--27`. Later on we'll see how to make unbreakable spaces of any length.

The ˜ has another important function: also says that this space should behave like a "normal" space, rather than a space after punctuation. TEX normally makes the space after a comma somewhat wider than a normal space, and the space after a period wider yet, following the traditional rules of typography.

Arbitrary horizontal spacing

To get a horizontal space (that is, a space between two words) as big as you want, type `\hskip` followed by a dimension:

`3\hskip 3pc 2\hskip 2pc 1\hskip 1pc 0` 3   2   1   0

The most common units for dimensions are inches (`in`), points (`pt` ; there are around 72 points in an inch), and picas (`pc` ; a pica is worth twelve points). And, for those who prefer to go metric, there are centimeters (`cm`) and millimeters (`mm`). Notice that there is no backslash before these units.

There is also a unit of horizontal space, the em, that depends on the current font. Traditionally, this was the width of an 'm', but in fact the two can be quite different: for example, for the font used here one em equals 10 pt, while an 'm' measures slightly less than 8 pt. This unit is useful if you want your spacing to be proportional to the size of the current font—in particular, when you're defining a command that should work with a variety of fonts.

Plain TEX has three predefined control sequences that generate this sort of proportional spacing:

- `\quad` corresponds to `\hskip 1em`;
- `\qquad` (a double quad) corresponds to `\quad\quad`;
- `\enskip` corresponds to half a quad.

If you use `\hskip` with a negative dimension, you get "negative spacing," that is, TEX backtracks and brings things closer together:

`AB, A\hskip -2pt B` .... AB, AB

We saw in section 4.6 that if you've specified a `\magnification`, all dimensions are multiplied by the magnification factor, except those whose units are preceded by the keyword `true`. For example, if you say `\hskip 1truein`, TEX will leave one inch of space in the output, no matter what the magnification.

5.2 Vertical spacing

Most of what we've said about horizontal spacing applies equally well to vertical spacing, that is, spacing between paragraphs. You get vertical spacing by typing \vskip followed by a dimension:

\vskip 5pt, \vskip 3mm, \vskip 4pc, \vskip -2pt.

The vertical counterpart of the em is the ex. An ex also depends on the current font; it is roughly the height of the letter 'x' (about 4.5 pt for this font).

Plain TEX has three predefined vertical skips:

- \smallskip skips 3 pt with an elasticity of plus or minus 1 pt (elasticities are explained below);
- \medskip skips 6 pt with an elasticity of plus or minus 2 pt;
- \bigskip skips 12 pt with an elasticity of plus or minus 4 pt.

5.3 Glue, or, Spaces that stretch and shrink

In practice, \hskip and \vskip are not sufficiently versatile to satisfy the requirements of page layout. To justify a paragraph, for instance—that is, to make all its lines the same length—it's necessary to stretch or shrink a bit the spacing between words, since only by the most unlikely of coincidences would the word widths add up exactly to the right amount. Pages, too, are often required to be of uniform height, and since each page can have many different elements, such as figures and equations, it would be hard to achieve uniformity if the spacing had to be exactly the same throughout.

TEX lets you add elastic spacing, informally known as glue, to your document. Glue stretches and shrinks (within predefined boundaries) as needed. To obtain glue, you use one of the normal spacing commands \hskip and \vskip, followed by three dimensions: the "ideal" amount of space you want to leave, the amount by which this ideal can be stretched, and the amount by which it can be shrunk. The stretchability and shrinkability are preceded by the keywords plus and minus (without a backslash). For example, if you say

\hskip 10pt plus 2pt minus 3pt

TEX will leave anywhere between 7 pt and 12 pt of space, depending on the constraints of the layout, and it will try its best to leave as close to 10 pt as it can. This ideal dimension is called the natural component of the glue. Either the plus or the minus part may be absent, but if both are present plus should precede minus:

\vskip 2in plus .5in,   \hskip .2em minus .05em.

Much of the glue on a page is put there automatically, without your having to think about it. For example, the spacing between words on this page is glue! In this font, it corresponds to \hskip 2.5pt plus 1.25pt minus 0.83pt. Imagine typing this expression by hand every time... Another common way to get glue is by using macros like \smallskip and its sisters (section 5.2).

How does TEX decide by how much each blob of glue must be stretched or shrunk? To understand this, we must know a bit about the way in which paragraphs and pages are built up. Let's look at paragraphs first. Roughly speaking, a paragraph is created in three stages:

• First, TEX sets the whole paragraph in a single line, as long as necessary. As it does this, only the natural component of the glue is considered, so `\hskip 10pt plus 2pt minus 3pt` counts as 10 pt.

• Then TEX breaks up this long line into several lines of length approximately `\hsize`, the page width. It generally tries several possibilities to find the best possible solution.

• If a line is too short, TEX stretches each blob of glue on it in proportion to its stretchability, till the line reaches the desired size. Thus, if you write `\hskip 6pt plus 2pt` and `\hskip 0pt plus 4pt` on the same line, the second blob of glue will stretch twice as fast as the first, even though its natural dimension is zero. Similarly, if a line is too long, TEX shrinks the spaces that occur in it in proportion to their declared shrinkability. In particular, a space that is declared without `plus` or `minus` never changes size, because it has no elasticity.

We said above that `\hskip 10pt plus 2pt minus 3pt` will produce between 7 pt and 12 pt of space, but that's not quite true. If a line is too short even after its stretchability has been added, TEX will overstretch it, and write a message like

```
! Underfull \hbox (badness 10000) detected at line 210
```

on your screen. (A line is a special case of an `\hbox`.) The badness of a line is a measure of how much it had to stretch or shrink to satisfy the constraints imposed on it. It is a relative measure: a line with more elasticity can stretch and shrink more than one with less elasticity, and yet get the same badness rating. If the badness is 10000 (the maximum), the glue has been overstretched.

On the other hand, if a line is too long even after its shrinkability has been taken into account, TEX won't overshrink it; it just makes it as short as the shrinkability allows, and sends you an `Overfull \hbox` message. It also prints a black stroke, of width `\overfullrule`, to the right of the line, like this:

This line is too long because TEX doesn't know how to hyphenate "manuscript." ▮

The elasticity of vertical glue is likewise used by TEX to make pages conform to a preset size. The process is very similar to the one for lines: TEX fills up more than a page's worth of text, then tries to find a suitable breakpoint. Once it finds it, it stretches or shrinks the vertical glue on the page in proportion to its elasticity.

Lines and pages are particular cases of horizontal and vertical boxes, as we'll see in chapter 8. When TEX builds up a box whose size is fixed beforehand, it uses the elasticity of the glue inside to meet the size requirement, just as it does when it justifies lines. In other words, if the natural width (for a horizontal box) or the natural height (for a vertical box) of the material inside the box is less than the box's target width or height, TEX stretches the glue inside the box to try to meet the target; and similarly if the width or height is insufficient.

An example

Plain T_EX's `\line` command takes its argument—the material that follows in braces—and makes with it a line of length exactly `\hsize`, stretching all the way between the left and right margins. You can imagine it as a groove inside which the characters slide. If we say `\line{A\hskip 60pt B\hskip 100pt C\hskip 40pt D}`, we get

A   B   C   D

and a complaint about an underfull box, because the length of the material in this line, 200 pt plus the widths of the letters, is only about 228 pt, versus the desired 327 pt of `\hsize`. There is a deficit of 99 pt. If instead we say

```
\line{A\hskip 60pt plus 100pt B\hskip 100pt
        C\hskip 40pt plus 50pt D}
```

we've got more than enough stretchability to cover the deficit, so T_EX no longer complains the line is underfull. The result is

A   B   C   D

The middle space didn't stretch, because we gave it no elasticity. The first space had twice as much stretchability as the last, so it stretched twice as much—since the deficit was 99 pt, the increments were $2/3 \times 99 = 66$ pt and 33 pt, respectively. (If this arithmetic makes you dizzy, don't worry—the important thing is that the stretching is proportional to the `plus` component of the glue, and the shrinking is proportional to the `minus` component.)

5.4 Springs

Glue is meant to stretch or shrink only to a certain point. T_EX also has springs, which can stretch indefinitely. *Springs don't create new spacing, they just fill up space created by other commands.* You can imagine that they are made of very thin wire; have you ever seen thin wire pierce through concrete?

T_EX has two predefined types of horizontal springs, to fill up horizontal space, and two types of vertical springs, to fill up vertical space. They're called `\hfil`, `\hfill`, `\vfil` and `\vfill`.

If we say `{\hfill XXX YYY}` or `{XXX\hfill YYY}` or `{XXX YYY \hfill}`, nothing happens: it's as if the `\hfill` weren't there. The group doesn't create any empty space, so the spring doesn't stretch. Now let's make up some empty space by forcing the group to fill up 1.2 inches:

```
\hbox to 1.2in{$|$\hfill XXX YYY$|$} .......... |      XXX YYY|
\hbox to 1.2in{$|$XXX\hfill YYY$|$} ........... |XXX          YYY|
\hbox to 1.2in{$|$XXX YYY\hfill$|$} ........... |XXX YYY          |
\hbox to 1.2in{$|$XXX\hfill YYY\hfill$|$} .... |XXX    YYY    |
```

The spring stretches to fill up all the available space.

Here's the same experiment with a \vfill inside a \vbox whose size we set in advance, to make up empty space:

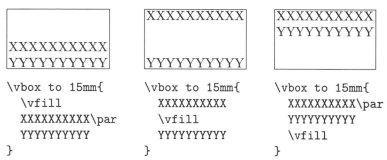

```
\vbox to 15mm{          \vbox to 15mm{          \vbox to 15mm{
   \vfill                  XXXXXXXXXX              XXXXXXXXXX\par
   XXXXXXXXXX\par          \vfill                  YYYYYYYYYY
   YYYYYYYYYY              YYYYYYYYYY              \vfill
}                       }                       }
```

Springs not only stretch indefinitely, but their stretching power is infinitely greater than that of normal glue. The result of

```
\line{A\hskip 0pt plus 1000pt B\hfil C\hskip 0pt plus 20pt D}
```

is that the glue between the first two letters and between the last two doesn't stretch at all, because the spring preempts their expansion:

AB   CD

Something similar happens if we mix the two types of horizontal springs:

```
\hbox to 25mm{$|$\hfil XXX\hfill YYY\hfil$|$} ... |XXX     YYY|
```

It's as if the \hfil didn't exist! A similar experiment with \vfil and \vfill inside the same \vbox would lead to the same conclusion. In other words, a spring with two 'l's (\hfill or \vfill) is infinitely stronger than its sibling with one 'l' (\hfil or \vfil). A stronger spring preempts any action on the part of a weak one.

On the other hand, if two or more \hfil or \vfil have to compete for the same space, they expand by the same amount. In section 7.5 you'll see this property in action, when we use vertical strings to automatically place a title. An even commoner example is centering: if you say \line{\hfil text\hfil}, you get the text exactly centered between the left and right margins.

Weak springs are consistently used inside macros: for example, \matrix and \over use weak springs to center their arguments. This gives you the chance of overriding their action with a strong spring; for example, the three fractions

$$\frac{a}{x_1 + \cdots + x_n}, \qquad \frac{a}{x_1 + \cdots + x_n}, \qquad \frac{a}{x_1 + \cdots + x_n}.$$

come from typing ${a\over x_1+\cdots+x_n}$, ${\hfill a\over...}$ and ${a\hfill\over...}$, respectively.

So the thing to do is to use strong springs in the text and weak springs when writing a macro.

5.5 Spacing and breaks

Suppose you've typed `...XXX\quad YYY...` and TₑX decides to start a new line between XXX and YYY. What will it do with the `\quad`? Should it leave it at the end of the line, or start the new line with it? Either way, the result would look strange. Instead, TₑX simply removes the space. More generally,

- TₑX loves spaces between words, as well as explicit `\hskip` commands, because they indicate allowable line breaks in a paragraph.
- When TₑX breaks lines at a space or `\hskip`, the space disappears.

But there are times when a break at an `\hskip` would be undesirable. For example, if you type `W\hskip -1pt A` inside a word, to achieve what typographers call better kerning, it would be disastrous to break the line between the two letters!

You can get a horizontal space at which TₑX will never break the line by typing `\kern` followed by a dimension:

<code>\kern 1em, \kern 3mm, \kern -.1pt</code>

In the example above, then, we could write `W\kern -1pt A`. Plain TₑX offers three predefined kerns:

- `\enspace` gives half an em;
- `\thinspace` gives one-sixth of an em (about 1.5 pt for this font);
- `\negthinspace` gives minus one-sixth of an em, that is, it pulls the surrounding letters together.

Warning: kerns have no elasticity! If you type `\kern 1em plus .2em` you get in the output a quad of space followed by the text plus .2em.

Now it can still happen that TₑX breaks a line right before a `\kern`, if there is an `\hskip` there. In this case TₑX will discard the kern as well as the `\hskip`. How then can you get a horizontal space that is guaranteed never to disappear, no matter what happens? Plain TₑX has a macro `\hglue` that you can use in this case:

<code>\hglue 2mm,   \hglue 5pt plus 2pt minus 2pt.</code>

All this stuff about horizontal spacing has a vertical counterpart. As we've seen, TₑX always typesets ahead a little bit, filling more than one page and then looking for a suitable place for a page break. Here again, a `\vskip` is a good target for a break; if the break happens there, the `\vskip` is eliminated, so it doesn't appear at the bottom of the page or at the top of the next.

To avoid having a page break at a vertical space, use `\kern` instead of `\vskip`. But if the `\kern` is next to a `\vskip` at the bottom or at the top of a page, it will disappear too.

To obtain a vertical space that TₑX can never throw out, use `\vglue`. For instance, to start your first page two inches from the top margin, say `\vglue 2in` at the top of your document; chapter 7 has more applications. With `\vglue`, as with `\hglue`, you can use `plus` and `minus`.

The `\kern` command is ambiguous: its effect depends on the current mode. In horizontal or math mode, `\kern` creates horizontal space, while in vertical mode, it creates vertical space. So if you conclude a paragraph with `...the end.\kern .5in`, nothing special will happen, since the kern will just be placed at the end of the last line, where there is room, or it will disappear at the break after the last line. But if you finish off the paragraph with `...the end.\par\kern .5in`, TₑX will put in half an inch of leading after the last line.

Similarly, if you type `\kern 5mm Start...` to get a 5 mm indentation at the beginning of a paragraph, you're in for a surprise: when TₑX encounters the `\kern`, it is in vertical mode, so it leaves 5 mm of leading above the paragraph. TₑX will only start horizontal mode when it reads the first character. In order to indent a paragraph by 5 mm, type instead `\hskip 5mm Start...`

5.6 Summary of basic spacing commands

There are three basic commands to get horizontal spacing of a desired width:

- `\hskip` (possibly followed by `plus` and/or `minus`). This is the most common type of horizontal spacing, or glue. TₑX can break the line there, in which case the glue disappears. When TₑX is in horizontal mode, a space in your input (or more than one, as we saw in section 5.1) is turned into an `\hskip` of the appropriate size. The springs `\hfil` and `\hfill` are special cases of `\hskip`: their natural width and shrinkability are zero, and their stretchability is infinite.

- `\kern` (in horizontal mode; no `plus` or `minus` allowed). TₑX will not break lines at a kern; but if it breaks just before or after it (at an `\hskip`, for example), the kern disappears. A ~ in the input is essentially equivalent to a `\kern`.

- `\hglue` (possibly followed by `plus` and/or `minus`). This cannot be discarded. In practice, this command is utilized seldom.

There are also three basic commands to get a desired amount of vertical spacing:

- `\vskip` (possibly followed by `plus` and/or `minus`). This is the most common type of vertical spacing, being used directly and through the macros `\bigskip`, `\medskip` and `\smallskip`. TₑX can break the page at a `\vskip`, in which case the `\vskip` disappears. The springs `\vfil` and `\vfill` are special cases of `\vskip`.

- `\kern` (in vertical mode; no `plus` or `minus` allowed). TₑX will not break pages at a vertical kern; but if it breaks just before or after it (at a `\vskip`, for example), the kern disappears. Vertical kernels are rarely used.

- `\vglue` (possibly followed by `plus` and/or `minus`). This cannot be discarded. This command is used mostly to leave space for figures and the like.

5.7 Spacing between paragraphs

The amount of vertical spacing between paragraphs is controlled by the variable `\parskip`. Plain TₑX sets it as follows: `\parskip=0pt plus 1pt`. You can increase it if you like, keeping it elastic to help the page layout:

`\parskip=4pt plus 2pt minus 2pt`

5.8 More springlike creatures

A spring can leave a trail as it stretches, filling the available space with copies of some material. A trail of dots, as in a table of contents, for example, is known as leaders; in T_EX this name is generalized to any visible spring.

Predefined leaders

There are several predefined leaders, all horizontal:

- \hrulefill draws a horizontal line, or rule;
- \dotfill prints a sequence of dots (the original leaders);
- \rightarrowfill makes an arrow pointing right;
- \leftarrowfill makes an arrow pointing left;
- \downbracefill makes horizontal braces opening down;
- \upbracefill : makes horizontal braces opening up.

The first four are used exactly like \hfill (with two 'l's), and have the same power as \hfill . So in the construction

```
\line{\leftarrowfill\ AAA \hrulefill\ BBB \dotfill\ MMM
\dotfill\ YYY\hrulefill\ ZZZ \rightarrowfill}
```

the available space is evenly distributed among all five "leaders:"

←——————— AAA ————— BBB . . . . . . MMM . . . . . . YYY —————— ZZZ ———————→

The remaining two predefined leaders, \downbracefill and \upbracefill , are slightly trickier to use because their height depends on the context. They are discussed again in chapter 9 and in the Dictionary; here we just show them by themselves:

\hbox to 2in{\downbracefill} . . . . . .

\hbox to 2in{\upbracefill} . . . .

. . . . To get the first line of this paragraph to end here we put a \break after the here ; to get it to start with dots from the left margin, we preceded it with \dotfill . But when we ran T_EX we got the message

```
! Leaders not followed by proper glue.
```

What happened? It turns out that horizontal leaders can only be used in horizontal mode, and we said \dotfill while T_EX was in vertical mode. To fix this, we used \leavevmode :

```
\leavevmode\dotfill To get the ... here\break
```

The \leavevmode can be replaced by \indent , \noindent , or any other command that pushes T_EX into horizontal mode (section 6.1). Ending a paragraph with dots to the right margin involves a different problem: if you add \dotfill\par to the last line, no dots will appear, because springs are discarded at the end of a paragraph. The solution is to fool T_EX into thinking there is some stuff after the leaders. We do this using \null , which makes an empty box: . . . .

```
... an empty box:\dotfill\null\par
```

5.9 Leaders in their full glory

All of the commands in the previous section are based on the `\leaders` control
sequence. This command is so versatile that it is worthwhile taking a closer look
at it. For example, in a table of contents, `\dotfill` looks too crowded; you can
get better results as follows:

```
\def\widedotfill{\leaders\hbox to 15pt{\hfil.\hfil}\hfill}
\parindent=0pt
Chapter 1. House plants \widedotfill 3 \par
Section 1.1. Harmless plants \widedotfill 15 \par
Section 1.2. Poisonous plants: what you should do in case
   of poisoning and what you absolutely must not do
   \widedotfill 37\par
```

These leaders are spaced by an amount that you can specify—here 15 pt—and
aligned vertically, which would not be the case with `\dotfill`. Notice the `\par`
commands, used to start a new line. (In practice you'd use blank lines, but here it
saves space.) The space created in this way is then filled with the leaders.[1]

You don't have to know in detail how `\leaders` works. All you need to know is
that you can replace the contents of the `\hbox` in the definition of `\widedotfill`
above by anything you want. Try the following constructions:

```
\hbox to .1in{\hfil$*$\hfil}
\hbox to 10mm{$\hfil\circ\hfil$}
\hbox{ \TeX\ }
```

We will come back to the TeX example at the end of the next section.

5.10 For the experienced user

The amount of glue inserted by TeX when it reads a space in horizontal mode
is a function of the current font, and is known from the corresponding `tfm` file.
However, you can change it by setting the variables `\spaceskip` (for normal
spacing) and `\xspaceskip` (for extra spacing after punctuation). For instance,
`\spaceskip=.3em \xspaceskip=.5em` makes the spacing between words from
there on completely inelastic (cf. the `\raggedright` macro of section 6.6), while

```
\spaceskip=.2em plus .2em minus .1em
\xspaceskip=.4em plus 1em minus 1em
```

---

[1] Actually, things are a bit more complicated. At the end of a paragraph TeX adds an amount of
white space given by the quantity `\parfillskip`, which plain TeX sets to a weak spring (in spite of
the two 'l's). This spring gets crushed to nothing when the same line contains the infinitely stronger
`\widedotfill`. See section 6.6 for details.

gives it a certain elasticity. These values will last only until the end of the current group, or until they're overridden by another assignment.

Negative springs

Once you have the notion of a spring firmly in mind, it is not hard to extend it to "negative springs," which have more shrinkability than their natural length. You can imagine them as being made of antimatter: instead of stretching, they contract and their length becomes negative! Consider, for example, the following definition:

$$\def\negspring{\hskip 0pt minus 1fil}$$

(As you may surmise, `fil` is the "unit" of weak springiness, and `fill` is its strong counterpart. `\hfil` is essentially the same as `\hskip 0pt plus 1fil`.)

You may be wondering what the use for `\negspring` is. But it's amazing what you can do with the beast, once you get the idea of how it works:

```
\vrule\hbox to 1pt{\negspring AAA ZZZ}\vrule . . . . . . . . . . . AAA ZZZ|
\vrule\hbox to 1pt{AAA ZZZ\negspring}\vrule . . . . . . . . . . |AAA ZZZ
```

The box in these constructions is delimited by vertical bars on either side for visibility, and its length is preset to almost zero. TEX cancels the length of the text AAA ZZZ with the shrinkability of the `\negspring`, with the result that the text comes out of the box on the side of the spring.

Plain TEX's `\rlap` and `\llap` macros use this idea. Saying `\rlap{...}` is like typesetting the stuff in braces and then backtracking as if you hadn't typeset anything; `\llap` is similar, but it backtracks first. As an application, we take another look at the table of contents from the previous section. Notice that the text there runs too close to the page numbers, impairing legibility. Here's an improvement:

```
\def\widedotfill{\leaders\hbox to 15pt{\hfil.\hfil}\hfill}
\def\page#1{\widedotfill\rlap{\hbox to 25pt{\hfill#1}}\par}
\rightskip=25pt
Chapter 1. House plants\page{3}
Section 1.1. Harmless plans\page{15}
Section 1.2. Poisonous plants: what you should do in case
    of poisoning and what you absolutely must not do\page{37}
```

Chapter 1. House plants 3
Section 1.1. Harmless plans 15
Section 1.2. Poisonous plants: what you should do in case of poisoning
and what you absolutely must not do 37

The trick here is to reduce the line length by 25 pt on the right, using `\rightskip` (section 6.5), and putting the page numbers past the end of the shortened lines using `\rlap`, which fools TEX into thinking the page number has width zero. Also, we have included the page number and the `\par` in the definition of the `\page` control sequence—you don't need to have read chapter 12 to figure out what's going on.

TEX also has a horizontal spring \hss and a vertical spring \vss, whose length can vary between $-\infty$ and $+\infty$. Like \hfil and its friends, \hss and \vss are primitives; but, if they weren't, we could define them like this:

```
\def\hss{\hskip 0pt plus 1fil minus 1fil}
\def\vss{\vskip 0pt plus 1fil minus 1fil}
```

The \centerline macro is equivalent to \line{\hss...\hss}. If the length of the text is no greater than \hsize, the spring \hss behaves like an ordinary spring and centers the text. On the other hand, if the text is too long for the page width, \hss contracts and allows the text to overflow by an equal amount on either side, without making TEX complain.

In section 9.5 we will meet another spring of negative length: \hidewidth.

5.11 Examples

Typesetting this chapter

Here are the various types of spaces used in this chapter. First of all, we suppressed the indentation at the beginning of paragraphs, and registered the fonts used for headings, once and for all:

```
\parindent=0pt
\font\chapnumfont=     HelveticaB at 35pt
\font\chaptitlefont=   HelveticaB at 22pt
\font\sectitlefont=    HelveticaB at 12pt
\font\subsectitlefont=HelveticaB at 10pt
```

(See more on indentation in the next chapter.) To get the chapter heading we typed

```
\hfill{\chapnumfont 5}
\medskip
\hfill{\chaptitlefont Spacing, glue and springs}
\vskip 14pc
```

Grouping limits the effect of the fonts to the title. Section headings come next. The previous section started with

```
\vskip 20pt plus 8pt minus 8t
\hskip-4.75pc
{\sectitlefont 5.9 For the experienced user}
\medskip
The amount of space ...
```

The \hskip-4.75pc makes the heading start about two centimeters, or .8 inches, to the left of the margin. Subsection headings are very similar:

```
\medbreak
{\subsectitlefont Negative springs}
\smallskip
Once you have ...
```

Every time we want a bit of space between two lines, we use a \smallskip. Finally, when we want to center a display, we type

\smallskip\centerline{...}\medskip

So here are all the nuts and bolts of this chapter's first page:

```
\input book.mac
\hfill{\chapnumfont 5}
\medskip
\hfill{\chaptitlefont Spacing, glue and springs}
\vskip 14pc

What sets apart a truly beautiful typesetting job
is the treatment of white space!  For this reason
\TeX\ has a rich set of commands devoted to the
control of spacing.

\vskip 20pt plus 8pt minus 8t
\hskip-4.75pc
{\sectitlefont 5.1 Horizontal spacing}
\medskip

{\subsectitlefont
The space bar and the carriage return key}
\smallskip

We saw in chapter 1 that \TeX\ gives special
treatment to spaces and carriage returns (represented by
{\eightrm SP} and {\eightrm CR}) in the sense that they
don't always appear in the printed output.
Here are the rules again:

{\parindent=3em\smallskip
\meti{$\bullet$}
Several consecutive ... printed document.
...
\meti{$\bullet$}
One or more spaces ... {\rm \TeX book}.
\smallskip}

This last rule...
```

The \input command, as explained in section 1.7, reads in a file containing style commands and macro definitions for this book. In addition, there are two commands that we haven't talked about before:

• The \meti command places its argument (here a bullet) in the paragraph indentation.

• The font command \eighttm brings in eight-point Times PostScript. You should be able to figure out its definition.

We will see in section 12.3 how a sequence of commands such as the ones used to open each chapter, can be encapsulated into a macro. In practice, we start a chapter with `\chapter{5}{Spacing, glue and springs}`, instead of explicitly typing four lines of spacing and font change commands. The two groups following `\chapter` are the macro's *arguments*; TEX plugs them in place of `#1` and `#2` in the macro definition:

```
\def\chapter#1#2{\vfil\eject
\hfill{\chapnumfont #1}
\medskip
\hfill{\chaptitlefont #2}
\vskip 14pc}
```

For other macros used for formatting this book, see `\section` and `\subsection` in the Dictionary. The use of macros has many advantages: it saves typing, ensures consistency from one chapter to the next, and makes the source file easier to understand.

The TEX logo

You may be curious about the definition of the TEX logo:

```
\def\TeX{T\kern-.1667em\lower.5ex\hbox{E}\kern-.125em X}
```

To lower the 'E', we use the `\lower` command, discussed in section 8.6. Notice how shifts are expressed in terms of `em` and `ex`, so they work correctly no matter what the current font is.

To obtain the pattern

TEX TEX TEX TEX TEX TEX TEX
TEX TEX TEX TEX TEX TEX TEX
TEX TEX TEX TEX TEX TEX TEX

we typed

```
\def\multitex{\leaders\hbox{\TeX\kern 1pt}\hfill}
\hskip 3cm\multitex\hskip 3cm\null\par
\hskip 3cm\multitex\hskip 3cm\null\par
\hskip 3cm\multitex\hskip 3cm\null\par
```

The `\null` prevents the `\hskip` from disappearing at the end of the line; see section 5.8. A vertically repeated pattern like this can also be obtained using vertical leaders; see the Dictionary under `\leaders`.

6
Paragraphs

6.1 Beginning and ending a paragraph

At the beginning of a job, TeX is in vertical mode. When a paragraph starts, TeX passes to horizontal mode. The end of a paragraph corresponds to a return to vertical mode. (These are the ordinary horizontal and vertical modes. There are also horizontal and vertical modes inside boxes, but we'll ignore them for now.)

If TeX is in vertical mode, it passes to horizontal mode when it encounters:

- a character;
- one of the control sequences \indent, \noindent, \leavevmode;
- a math formula, delimited by dollar signs $ (between the dollar signs TeX is in math mode, but after that it goes into horizontal mode);
- any command that makes sense only in horizontal mode, such as \hskip, \vrule, or one of the paragraph formatting commands to be discussed later.

While in horizontal mode, TeX switches to vertical mode, completing the current paragraph, when it encounters:

- vertical spacing commands such as \vskip, \smallskip, \medskip and \bigskip, or their variants \smallbreak, \medbreak and \bigbreak;
- the \par command or its alias, a blank line (that is, two or more consecutive CR characters);
- any command that makes sense only in vertical mode, such as \hrule.

6.2 What's in a paragraph?

A paragraph generally contains text, that is, characters one after another. But you can also put inside a paragraph a box or a rule (which must be called a `\vrule`).

Recall from section 5.3 that T_EX reads a whole paragraph before trying to typeset it. Then it creates a very long line, without worrying about the width of the page. Next it tries to find line breaks: first between words, or, if that doesn't work, between syllables. Once the breakpoints are determined, T_EX stretches or shrinks the glue in each line so that they come out with the same length `\hsize` . It there's no way to do this, it sends the user an error message, such as `Overfull \hbox`, and draws a vertical stroke on the margin, next to the offending line. Finally, it stacks up the lines.

6.3 Automatic indentation

The first line of a paragraph is generally indented. For this book, this feature was turned off, but we turned it back on at the beginning of this paragraph. The amount of indentation is given by `\parindent` , and indentation is turned off by setting this to zero. Plain T_EX defines `\parindent` as follows:

parindent=20pt

You can change this value. But watch out: if your document file starts by inputting a style file, like the example in section 5.11, which started with `\input book.mac` , any changes to the `\parindent` should come after the `\input` . This is because the style file most likely resets the `\parindent` .

Indentation can also be negative, meaning that the first line starts to the left of the margin. This is what happened in this paragraph: we typed `\parindent=-.5in` just before it. You'll find another example of negative indentation at the end of section 6.5.

If you want only one paragraph to start without indentation, you can precede it by `\noindent` . This control sequence switches T_EX from vertical to horizontal mode.

6.4 Obeying lines

As you know, T_EX generally ignores carriage returns, or rather, treats them as spaces. But there are cases when it is desirable to have T_EX respect the line breaks of the input file. To achieve this, you should start your text with the command `\obeylines` , which makes CR equivalent to `\par` :

```
\obeylines
Old pond
The sound of a frog %
  jumping in the water
Is heard.
\smallskip
Matsuo Basho
```

Old pond
The sound of a frog jumping in the water
Is heard.

Matsuo Basho

This example shows that a long input line can be broken, with a % at the end of the first half to hide the CR. It also shows that to get any sort of vertical spacing you still need explicit commands: several CRs in a row, with blank lines in between, have exactly the same effect as one.

As there is no simple command to counteract the effect of \obeylines, you should enclose in braces the region where it should have effect. Here's a common example, which you can imitate when writing a letter:

```
{\obeylines
\hfill November 9, 1989
\medskip
Raymond Seroul
UER de Math\'ematiques et d'Informatique
7, rue Ren\'e Descartes
67000 Strasbourg, France}
```

The result is:

November 9, 1989

Raymond Seroul
UER de Mathématiques et d'Informatique
7, rue René Descartes
67000 Strasbourg, France

6.5 Left and right margins

{\leftskip=.5in The variables \leftskip and \rightskip control the relative position of the left and right margins. This paragraph was preceded by \leftskip=.5in, so its left margin was moved in (to the right) by .5 in. Had it been \rightskip=.5in, the right margin would have been moved in (to the left) by the same amount. Had the dimension been negative, the change would have been in the opposite direction.

To contain the effect of the change in \leftskip to a portion of the text, we used braces to start a new group. As soon as T_EX reads the matching braces at the end of the next paragraph, \leftskip will revert to its old value of zero.

So why is this paragraph not pushed in like the previous two? Because it's not quite finished when T_EX reads the braces and restores the \leftskip! *The values of* \rightskip *and* \leftskip *applied to a paragraph are those in effect when the paragraph ends.* }

In order for the last paragraph in a group to be affected by a change in \rightskip or \leftskip that is local to the group, it is necessary to have \par (or a blank line, or some such) before the right brace:

```
{\leftskip=.5in
The variables ...
... when the paragraph ends.
\par}
```

Of course, the same precaution must be taken when \rightskip and \leftskip are changed by the action of some command, as is the case with many macros introduced in this chapter.

There's another little trap related to the one we've just discussed. By now you know better than to fall into it, but here's an example anyway:

```
first paragraph                   first paragraph
\leftskip=1cm
                                  \leftskip=1cm
second paragraph                  second paragraph
```

In the example on the left, both paragraphs are pushed in by 1 cm; in the example on the right, only the second paragraph is so indented.

Hanging indentation

One can combine indentation with \leftskip to get hanging indentation, the
effect displayed in this paragraph. Here we typed {\parindent=-1cm and
\leftskip=1cm...\par}; in section 6.9 we'll discuss special commands
that can be used to achieve the same effect.

6.6 Ragged margins

Justification is fine and good, but every now and then one wants ragged margins, especially when setting text in a narrow column. How can TeX be stopped from justifying lines? To understand the solution to this problem, let's take another look at the variables \leftskip and \rightskip which govern the left and right margin offsets. The truth is, they don't change the margins at all! When you type \leftskip=1cm, TeX inserts 1 cm of white space at the beginning of each line; this gives the impression that the left margin is pushed in.

The usefulness of this behavior lies in that \rightskip and \leftskip don't have to be fixed amounts. For example, by giving \rightskip a bit of stretchability, we tell TeX that it can leave some white space at the end of each line:

The first of these was to accept nothing as true which I did not clearly recognize to be so: that is to say, carefully to avoid precipitation and prejudice in judgements, and to accept in them nothing more than what was presented to my mind so clearly and distinctly that I could have no occasion to doubt it.[1]

To format this paragraph, we used plain TeX's \raggedright macro, like this: {\raggedright The first ... doubt it.\par} Here is the definition of \raggedright:

```
\def\raggedright{\rightskip=0pt plus 2em
    \spaceskip=.3333em \xspaceskip=.5em}
```

---

[1] This and the next three quotations form Descartes's four principles (*Discours de la méthode*, part II, translated by Haldane and Ross).

The idea is to give stretchability to \rightskip, while at the same time taking away the elasticity from interword spacing (where it is no longer necessary). As explained above, when you using a group to limit the scope of \raggedright, you must type \par before the end of the group; otherwise, the last paragraph won't be affected.

Plain TEX doesn't provide a \raggedleft macro: let's design one ourselves. We can imitate the definition of \raggedright, this time giving stretchability to \leftskip:

```
\def\raggedleft{\leftskip=0pt plus 2em
\spaceskip=.3333em\xspaceskip=.5em}
```

But when we try {\raggedleft The second ... \par}, something doesn't come out quite right:

The second was to divide up each of the difficulties which I examined into
    as many parts as possible, and as seemed requisite in order that it might
be resolved in the best manner possible.

What happened to the last line? Remember from section 5.3 that TEX places the lines of a paragraph in horizontal boxes (lines) of length \hsize —except for the last, which is generally shorter. TEX handles the special case of the last line by a trick of sorts: it automatically adds at the end of the paragraph a weak spring, essentially equivalent to \hfil. In this way the last line is treated just like the others—it gets stretched to length \hsize, but all the slack is taken up by this sneaky spring.

This worked well for justified and ragged right text, but it's not what we want here. Fortunately, this end-of-paragraph glue is not written in stone: it's just one more of TEX's variables, called \parfillskip. Plain TEX sets it to 0pt plus 1fil; if we make it zero, the last line will end at the right margin like the others. So we add \parfillskip=0pt to the definition of \raggedleft, and try again:

The second was to divide up each of the difficulties which I examined into
    as many parts as possible, and as seemed requisite in order that it might
                                            be resolved in the best manner possible.

This still leaves something to be desired: the last line is much shorter than the others, and in fact TEX declares it underfull, since the only stretchability in it comes from the \leftskip. To balance out the lines \leftskip must be given a lot more stretchability—in fact it must be allowed to stretch across the whole page (think of a paragraph containing a single word):

```
\def\raggedleft{\leftskip=0pt plus \hsize
  \parfillskip=0pt\spaceskip=.3333em\xspaceskip=.5em}
```

The second was to divide up each of the difficulties which I
examined into as many parts as possible, and as seemed requisite
in order that it might be resolved in the best manner possible.

6.7 Quotations

The `\narrower` command increases both `\leftskip` and `\rightskip` by an amount equal to `\parindent`. In other words, the left and right margins both move in. This is often useful for quotations:

> The third was to carry on my reflections in due order, commencing with objects that were the most simple and easy to understand, in order to rise little by little, or by degrees, to knowledge of the most complex, assuming an order, even if a fictitious one, among those which to not follow a natural sequence relatively to one another.

The code here was `{\parindent=.5in \narrower The third ... \par}`. As you probably have figured out, if you use `\narrower` within a group you must end the last paragraph before closing the group.

The first line of a paragraph to which `\narrower` applies normally receives a double indentation—the normal first-indentation, plus the `\leftskip`. If you don't want that, start the paragraph with `\noindent`.

If you type a second `\narrower` while the first is still active, their effects accumulate, because the change in `\rightskip` and `\leftskip` is relative. By contrast, `\raggedright` causes an absolute change in `\rightskip`. It follows that `\raggedright\narrower` works nicely, but `\narrower\raggedright` doesn't do what you might expect (try it out).

6.8 Centering text

To center text, you have two options. You can give `\leftskip` and `\rightskip` the same stretchability, and suppress the end-of-paragraph glue automatically added by TeX:

```
\leftskip=0pt plus .5in \rightskip=0pt plus .5in
\parfillskip=0pt
```

This leaves to TeX the task of figuring out where to break lines. The second solution lets you control where the line breaks go, using `\obeylines`. In this case it's best to make `\leftskip` and `\rightskip` into springs, that is, give them infinite stretchability:

```
{\leftskip=0pt plus 1fil
\rightskip=0pt plus 1fil
\parfillskip=0pt
\obeylines
The last was in all cases
...
... omitted nothing.\par}
```

> The last was in all cases
> to make enumerations so complete
> and reviews so general
> that I should be certain
> of having omitted nothing.

6.9 Series of items

An important use of hanging indentation is in formatting series of items, or enumerations. The \item command of plain TEX provides an easy way to do this.

1978 **Classic**—Some say it's the best red Bordeaux of this vintage. Others, that it is Margaux's finest wine since its 1961. It sells these days from $30 to $60.

1979 **Near-Classic**—Some critics have said that this is the best red Bordeaux of this vintage. It's big and rich, sells for around $60.

1980 ****—Selling for around $32.[2]

```
{\parindent=1cm
\item{1978} {\bf Classic}--- ... \$60.
\item{1979} {\bf Near-Classic}--- ... \$60.
\item{1980} ****---Selling for around \$32.
\par}
```

As you can see, \item starts a new paragraph and affects only that paragraph. It temporarily increases \leftskip by an amount equal to \parindent, thus moving in the left margin. It also places its argument—the contents of the braces following it—on the new margin. When the paragraph ends, the previous value of \leftskip is restored.

Since \item expects an argument, you should follow it with a group even if you have nothing to write on the margin: \item{}... If you don't do this, TEX, following its general rules for macro arguments (chapter 12), will look for the first character of the paragraph and use that as an argument. Give it a try, it won't hurt.

Although its effect is local anyway, \item is often used inside a group, as in the example above, so the amount of hanging indentation can be controlled by a local change in \parindent. In this case, as usual, the group must end with \par, otherwise the last paragraph is not handled correctly. (There is no need for \par between the items, because \item itself starts a new paragraph.)

The \itemitem macro is used exactly the same way as \item, and has the same effect, except that the left margin is pushed in by 2 \parindent:

1. To accept nothing as true which I did not clearly recognize to be so: that is to say,
   a. carefully to avoid precipitation and prejudice in judgements;
   b. to accept in them nothing more than what was presented to my mind so clearly and distinctly that I could have no occasion to doubt it.

```
{\parindent=20pt
\item{{\bf 1.}} To accept nothing as true which ...
{\itemitem{\bf a.} carefully to avoid precipitation ...
{\itemitem{\bf b.} to accept in them nothing more ...
\par}
```

---

[2] E. Frank Henriques, *The Signet Encyclopedia of Wine* (1984).

Plain TeX doesn't have an `\itemitemitem` macro. If you really need such a thing, you'll find it at the end of the chapter. Another possible macro for series of items is also defined there.

One often wants a page layout like this:

TeX software Maria Code, DP Services
1371 Sydney Drive
Sunnyvale, CA 94087

TeX support TeX USERS GROUP
P. O. Box 9506
Providence, RI 02940

The trick here is to get `\item` to set its tag flush left, rather than in its normal position, which is within .5 em of the indented left margin. To do this we defined a new macro, `\leftitem`, whose argument is the tag:

```
\def\leftitem#1{\item{\hbox to\parindent{\enspace#1\hfill}}}
```

This macro passes to `\item` a box of width `\parindent`. The tag starts .5 em to the right of the left edge of the box (why?); but the right edge of the box is placed by `\item` .5 em to the left of the paragraph margin, so everything cancels out and the tag is set flush against the outer margin! Here's how `\leftitem` was used:

```
\parindent=1in
\leftitem{\boldhelvetica \TeX\ software}
Maria Code, DP Services\hfill\break
1371 Sydney Drive\hfill\break
Sunnyvale, CA 94087
```

Notice the use of `\hfill\break` to terminate the lines; `\obeylines` wouldn't have worked, because it effectively makes a carriage return equivalent to a `\par`, and here we need to keep everything in the same paragraph.

The code above assumes that the tag is no longer than `\parindent` minus .5 em. If you have a very long tag, you may want to break it into lines, and set it in a `\vtop`, a type of vertical box:

TeX support TeX USERS GROUP
P. O. Box 9506
Providence, RI 02940

```
\parindent=.7in
\leftitem{\boldhelvetica
  \smash{\vtop{
    \hbox{\TeX}
    \hbox{support}}}}
\TeX USERS GROUP...
```

The `\cornerbox` macro of section 8.8, gives another way to achieve a similar effect.

6.10 More on hanging indentation

The two commands \hangindent and \hangafter work in conjunction. The first of them is followed by a positive or negative dimension, and the second by a positive or negative integer. This example illustrates better than any explanation the action of \hangindent and \hangafter in the various cases:

VAUX-LE-VICOMTE, 46 km southeast of Paris, is one of the great classical châteaux. Louis XIV's finance superintendent, Nicholas Fouquet, had it built at colossal expense using the top designers of the day—the royal architect Le Vau, the painter Le Brun and Le Nôtre, the landscape gardener.[3]

\hangindent=27pt
\hangafter=3

VAUX-LE-VICOMTE, 46 km southeast of Paris, is one of the great classical châteaux. Louis XIV's finance superintendent, Nicholas Fouquet, had it built at colossal expense using the top designers of the day—the royal architect Le Vau, the painter Le Brun and Le Nôtre, the landscape gardener.

\hangindent=-27pt
\hangafter=3

VAUX-LE-VICOMTE, 46 km south-east of Paris, is one of the great classical châteaux. Louis XIV's finance superintendent, Nicholas Fouquet, had it built at colossal expense using the top designers of the day—the royal architect Le Vau, the painter Le Brun and Le Nôtre, the landscape gardener.

\hangindent=27pt
\hangafter=-3

VAUX-LE-VICOMTE, 46 km south-east of Paris, is one of the great classical châteaux. Louis XIV's finance superintendent, Nicholas Fouquet, had it built at colossal expense using the top designers of the day—the royal architect Le Vau, the painter Le Brun and Le Nôtre, the landscape gardener.

\hangindent=-27pt
\hangafter=-3

Normally, \hangindent and \hangafter are placed at the beginning of the paragraph to which they apply, but they have the same effect if placed anywhere before the paragraph ends. As usual, the value used is the one in effect at the end of the paragraph. If \hangafter is not set, TEX assumes it to be 1.

6.11 Paragraphs with fancy shapes

The Count of Charolais had left the King a magnificent defensive position. The ridge of Montlhéry, running roughly west-east, rose steeply from the Paris road, came to a peak where stood the castle, and then declined eastward into the plain. A little to the west of the castle, on the northern versant, huddled the village of Montlhéry. Pierre de Brezé had marshaled his Norman gentry and squadrons of lances, all mounted, behind "a great ditch and hedge" at the bottom of the slope, facing the much superior numbers of the Count of St. Pol.[4]

---

[3] Kate Baillie and Tim Salmon, *The Rough Guide to Paris*.

[4] Paul Murray Kendall, *Louis XI* (1971).

This funnel-shaped paragraph started with

```
\eightpoint
\parshape=7 0cm 11.5cm .5cm 10.5cm 1cm 9.5cm 1.5cm 8.5cm
2cm 7.5cm 2.5cm 6.5cm 3cm 5.5cm\noindent The Count ...
```

The command `\parshape=7` says that the first seven lines of this paragraph should be treated specially. The first line is to be indented by 0 cm and its length should be 11.5 cm, the second line should be indented by 0.5 cm and its length should be 10.5 cm, and so on. Finally, the seventh line should be indented by 3 cm and be 5.5 cm long. If, as is the case here, the paragraph has more than the specified number of lines, TeX keeps repeating the specifications for the last one.

On page 101 of *The TeXbook* you will find some even more spectacular examples, such as two paragraphs that fit inside each other, one being in the shape of a circle and the other having a half-circle cut away. They are both formatted with `\parshape` and stored in boxes, then placed side by side with a negative `\hskip` to bring them together (cf. section 8.10).

An unpleasant surprise

The `\parshape` command acts on the paragraph in which it occurs and on that paragraph alone. It can be placed anywhere in the paragraph, even at the end, just like `\hangindent` and `\hangafter`. You're familiar with the reason by now. (The only shape control commands that are different are `\item` and `\itemitem`, because they start new paragraphs.)

The unpleasant surprise comes when you mistakenly exchange two lines. Imagine you want to format the second of two paragraphs using `\parshape`, as in the example on the left:

```
first paragraph                first paragraph
                               \parshape ...
\parshape ...
second paragraph               second paragraph
```

If you accidentally exchange the `\parshape` line with the empty line in your text editor, you end up with the code on the right. The result is that it's the first paragraph that is formatted, rather than the second! (Even though we're on guard against this sort of thing, this has happened to us several times.)

6.12 Footnotes

Plain TeX's `\footnote` macro takes two arguments: the number or symbol used to mark the note, and the text of the note. To get

Si Dieu nous a fait à son image, nous le lui avons bien rendu.*

Voltaire

---

* If God made us in his image, we have certainly returned the compliment.

we typed

```
Si Dieu nous a fait ... bien rendu.
\footnote{*}{If God made us ... the compliment.}
\smallskip \hskip 2in Voltaire
```

If you want your footnotes numbered sequentially, you can do it automatically with the following macro, which calls \footnote with the appropriate first argument:

```
\newcount\notenumber \notenumber=1
\def\myfootnote#1{\unskip\footnote{$^{\the\notenumber}$}{#1}%
  \global\advance\notenumber by 1}
```

Now you only need one argument, the text: \myfootnote{...}. The \unskip primitive does what its name says: it removes the last bit of glue in the current paragraph. Here it counteracts any spurious spaces that you may have inadvertently typed before the \footnote, and which would appear before the raised note number. [5]

You can also have \myfootnote use a different font for the note (by default, \footnote uses they same font for note and text, as in the example above): just precede the second occurrence of #1 in the code above by the appropriate font command. For an even more sophisticated version of \myfootnote, see the Dictionary.

6.13 Two new macros for the aspiring wizard

A triple item macro

Let's try to write an \itemitemitem macro to allow the creation of third-level lists. We can base ourselves on \item and \itemitem, whose definitions are given on page 355 of *The T_EXbook*:

```
\def\item{\par \hangindent=\parindent \textindent}
\def\itemitem{\par\indent \hangindent=2\parindent \textindent}
```

According to section 6.9, \hangindent=\parindent causes all lines in the current paragraph, starting with the second (because \hangafter=1), to be indented by \parindent. The command \hangindent=2\parindent works the same way. What \textindent does is not so clear, but we don't really need to know it—we can just extrapolate by adding one more \indent and replacing the 2 by 3 :

```
\def\itemitemitem{\par\indent\indent
    \hangindent=3\parindent \textindent}
```

This turns out to work perfectly. Here are all three item macros in action: the first paragraph (in bold) starts with \noindent, the second with \item{A.}, the third with \itemitem{1.} and the last two with \itemitemitem{...}. The \parindent is plain T_EX's default (20 pt).

---

[5] Like this.

South and Southeast Asia: the Late Colonial Period and the Emergence of New Nations Since 1920

A. India, Pakistan, Bangladesh, Ceylon, Tibet, and Nepal since 1920

    1. India since *c.* 1920: nationalism and the decline of the raj

        a. Dyarchy and the conflict between British policy and the aims of Indian nationalism: the Congress and Gandhi's technique of active, nonviolent revolution; Round Table Conferences

        b. The Government of India Act (1935), the political and economic effects of World War II, partition and independence (1947), Hindu–Muslim polarization[6]

An alternative for \item

The \item and \itemitem macros are very nice, but they have a problem: they're wasteful of space. Here is an alternative, which is after a fashion the opposite of \item. It indents the first line only, and puts its argument in the indentation. You can adjust the \parindent depending on how wide the label is.

```
\def\meti#1{\par\indent\llap{#1\enspace}\ignorespaces}
```

Each of the following paragraphs starts with \meti{\bf ...}; the \parindent is .5 in. We also opened things up a bit by inserting a \smallskip before each \meti.

    (i) If a sequence of random variables converges almost surely it converges in probability. The converse is generally not true.

    (ii) A necessary and sufficient condition for convergence in probability of a sequence of random variables is that, for any $\epsilon > 0$ and $\delta > 0$, there exists $n_0 = n(\epsilon, \delta)$ such that for $n, n' > n_0$ we have $P\{|\xi_{n'} - \xi_n| > \epsilon\} < \delta$.

    (iii) If a sequence of random variables converges in probability to ξ and also to η, we have $\xi = \eta \pmod{P}$.

It's easy to figure out how \meti works: the first line is indented by \parindent, then \llap writes to the left of the current point the stuff in braces, without moving the current point. As for \ignorespaces, it's something of a converse for the \unskip control sequence used in section 6.12: it makes T$_{\!E}$X ignore any spaces or carriage returns that follow the macro, so that \meti{...}XXX and \meti{...} XXX have the same effect.

---

[6] *Encyclopedia Britannica* (1988): *Propaedia.*

7
Page layout

7.1 Page layout in plain TEX

Plain TEX formats the output page like this:

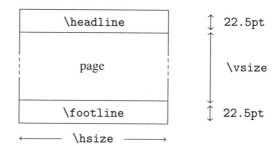

It sets the running headline and the footline as follows:

```
\headline={\hfil}
\footline={\hfil\tenrm\folio\hfil}
```

Notice the syntax: \headline and \footline are not macros, but *variables*; they're set with an equals sign = followed by a group in braces, rather than with \def . (The equals sign is optional.) The \folio command prints the page number. According to the code above, then, the running headline is blank, while the footline contains the page number, centered, in 10 point roman.

If you want no page numbers at all, start your document with \nopagenumbers ; this is an abbreviation for \footline={\hfil}. To get the page number at the top instead of at the bottom, switch the contents of \headline and \footline :

```
\headline={\hfil\tenrm\folio\hfil}
\footline={\hfil}
```

Printing the page number

The page number is stored in the variable \pageno, and increases automatically. Unless you say otherwise, the first page is numbered 1. In order for the first page to be 12, say, start your file with \pageno=12.

By convention, \folio prints a negative \pageno in roman numerals. If you want the page number in roman numerals, then set the \pageno to the corresponding negative number: \pageno=-12. (You may be wondering if the next page number will be. Don't worry, \pageno increases in absolute value.)

You can change \folio, if you wish, to use a different convention. See the Dictionary for a version that prints uppercase roman numerals.

7.2 A more elaborate layout

The layout provided by plain TEX is very stark. It does not distinguish between right and left pages, nor between the first page (which generally shouldn't carry a folio number) and the others. For this reason we list here the code for a more elaborate page layout, which you should put into a file `fancy.tex` for later use; the flexibility gained will be worth the effort.

```
\newif\iftitlepage    \titlepagetrue
\newtoks\titlepagehead \titlepagehead={\hfil}
\newtoks\titlepagefoot \titlepagefoot={\hfil}

\newtoks\runningauthor \runningauthor={\hfil}
\newtoks\runningtitle  \runningtitle={\hfil}

\newtoks\evenpagehead  \newtoks\oddpagehead
\evenpagehead={\hfil\the\runningauthor\hfil}
\oddpagehead={\hfil\the\runningtitle\hfil}

\newtoks\evenpagefoot \evenpagefoot={\hfil\tenrm\folio\hfil}
\newtoks\oddpagefoot  \oddpagefoot={\hfil\tenrm\folio\hfil}

\headline={\iftitlepage\the\titlepagehead
\else\ifodd\pageno\the\oddpagehead
\else\the\evenpagehead\fi\fi}

\footline={\iftitlepage\the\titlepagefoot
\global\titlepagefalse
\else\ifodd\pageno\the\oddpagefoot
\else\the\evenpagefoot\fi\fi}

\def\nopagenumbers{\def\folio{\hfil}}
```

None of this is too hard to understand; see section 12.8 for details. Perhaps the only non-obvious command is \the: it corresponds roughly to `write` in Pascal or `print` in Basic, effectively passing to the output stream the contents of a variable.

Throughout this chapter we'll assume that every document starts with the command

```
\input fancy
```

which tells TEX to read and keep the definitions in the file you just typed in. We will call the aggregate of definitions in this file the `fancy` format.

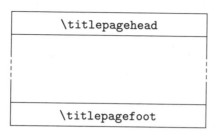

The diagram on the left represents the title page provided by the `fancy` format. The diagram below shows typical left and right pages. The control sequences shown in the diagrams are variables that govern the headlines and footlines of the various pages. You can fill them in with whatever you want; let's look at their default settings.

\evenpagehead	\oddpagehead
\evenpagefoot	\oddpagefoot

Default settings

Unless you say otherwise, the `fancy` format leaves the top and bottom of the title page blank. Other pages get a running headline with \runningauthor (if even) and \runningtitle (if odd), in the center; you can set these variables to the author and title of the document, or to title of book and chapter, and so on. The page number goes at the bottom, centered, in 10 point roman. If you put at the top of your file (right after \input fancy) the commands

$$\text{\textbackslash runningauthor=\{\textbackslash eightbf Raymond Chandler\}}$$
$$\text{\textbackslash runningtitle=\{\textbackslash eightbf Playback\}}$$

the pages following the title page will look like this:

Raymond Chandler	Playback
2	3

It is always necessary to specify a font inside the various headline font in head-lineand footline variables, either directly or, as in this example, by including the choice in the definition of \runningauthor and \runningtitle. The reason is that there is no way to tell what TEX will be in the middle of when it stops to break a page and typeset its headline and footline. If you don't set a font explicitly, TEX will use the current one, with results that can range from the humorous to the positively disastrous.

Other choices

If you don't want the first page to be a title page, perhaps because you're printing only one chapter of a book, just say `\titlepagefalse` right after `\input fancy`.

If you want a headline on the title page as well, you must set `\titlepagehead`. Here's one possibility:

<table>
<tr><td>R. Chandler: Playback</td><td>

```
\titlepagehead={\hfill
\eightbf R. Chandler:
Playback\hfill}
```

</td></tr>
</table>

A common choice for the running head of regular pages is this:

| 156 | R. Chandler | Playback | 157 |

in which case the footline is empty. You get this with

```
\evenpagehead={\tenbf\folio\hfill\eightrm\the\runningauthor}
\oddpagehead={\tenrm\the\runningtitle\hfill\tenbf\folio}
\evenpagefoot={\hfil} \oddpagefoot={\hfil}
\runningauthor={R. Chandler}
\runningtitle={Playback}
```

The next setup places page numbers on the margin, as in *The TEXbook*, creating an extended headline:

```
\evenpagehead={\llap{\tenbf\folio\quad}\eightit
  \the\runningauthor\hfill}
\oddpagehead={\hfill\eightit\the\runningtitle
  \rlap{\quad\tenbf\folio}}
```

And this runs a rule underneath the headline:

```
\evenpagehead={\vbox{\line{\tenbf\folio
  \hfill\eightrm\the\runningauthor}\smallskip\hrule}}
\oddpagehead={\vbox{\line{\eightrm\the\runningtitle
  \hfill\tenbf\folio}\smallskip\hrule}}
```

The footline, too, lends itself to many variations. Here is one, with the page number centered at the bottom, and a rule on each side:

```
\evenpagefoot={\hrulefill\quad\tenrm\folio\quad\hrulefill}
\oddpagefoot=\evenpagefoot
```

7.3 The title page

Université Louis Pasteur   Strasbourg, August 7, 1988
Laboratoire de Typographie
Informatique

A new laser printer driver

Setting up a head like this is similar to the problem of beginning a letter: the idea is to stack lines that are "paragraphs" by themselves. Here is an alternative to the approach given in section 6.4; it uses the `\line` command, which, as we saw in section 5.3, creates a horizontal box spanning the page. Using `\line` and springs, it is easy to get the layout above:

```
\line{Universit\'e Louis Pasteur\hfill Strasbourg, \today}
\line{Laboratoire de Typographie\hfill}
\line{Informatique\hfill}
\vskip 8pc plus 1pc minus 1pc
\line{\hfill\helveticabf A new laser printer driver\hfill}
```

These constructions are so common that plain T_EX offers abbreviations for them:

- `\centerline{...}` stands for `\line{\hfil...\hfil}` (actually, it has `\hss` instead of `\hfill`, but the difference is not important here);
- `\rightline{...}` stands for `\line{\hfil...}`;
- `\leftline{...}` stands for `\line{...\hfil}`.

Here is the counterpart of the code at the end of section 6.4 using `\rightline` and `\leftline`:

```
\rightline{November 9, 1989}
\leftline{Raymond Seroul}
\leftline{UER de Math\'ematiques et d'Informatique}
\leftline{7, rue Ren\'e Descartes}
\leftline{67000 Strasbourg, France}
```

Starting in midpage

Suppose you want to start your first page with a title two inches below the top margin. If you naïvely type

```
\vskip 2in\centerline{EXOTIC BUTTERFLIES}
```

you're in for a surprise: T_EX completely ignores the `\vskip`! This apparently obnoxious behavior follows from the rules discussed in sections 5.3 and 5.4, and is in fact entirely justified: T_EX is programmed to discard vertical spacing "between pages," just as it discards interword spaces at line boundaries. In order for a `\vskip` at the top of a page to be effective, *it must be preceded by something else*.

One often uses an empty box `\hbox{}` for this purpose; this trick is so common that the empty box has a name, `\null`. So you can start with

`\null\vskip 2in\centerline{EXOTIC BUTTERFLIES}`

and it will all work out. Another possibility is to replace `\vskip` by `\vglue` which, as explained in section 5.3, is guaranteed to leave glue that does not disappear. (It turns out that `\vglue` itself is a plain TeX macro based on the trick of using an invisible box, so the two solutions are actually one and the same.)

7.4 Starting a fresh page and leaving a blank page

Here's a variation on the same ideas: suppose you've filled half a page and you want to leave the other half blank. The command `\eject` tells TeX that it should start a fresh page. But if you just say `\eject`, you don't get the bottom half of the page blank! Instead the glue between lines and paragraphs in the top half get stretched so the text occupies the whole page. To change that behavior, you must put before the `\eject` a vertical spring `\vfill` to take up the slack.

Now suppose you want to leave a page blank, to insert a figure, for example. The naïve solution, `...\vfill\eject\vfill\eject`, won't work: the reason is again disappearing glue. When the second `\vfill\eject` comes along, TeX is already at the top of a page, so it ignores the `\vfill` and the `\eject` has no effect (`\eject` effectively means "this is an obligatory break," so it has no effect at a point where there is a break already). The correct solution again uses an empty box:

`...\vfill\eject\null\vfill\eject`

7.5 Placing a title

Since springs of the same power share evenly the available space, you can place a title, say, two-thirds of the way between the top margin of the page (or the document head) and the beginning of the text:

FAIRY TALES

Once upon a time...

```
\null
\vfill\vfill
\centerline
  {\bf FAIRY TALES}
\vfill
Once upon a time ...
\eject
```

The `\eject` is necessary to indicate where the page must end; if it isn't there TeX will fit as much text on the page as there is room for, and there will no space for the springs to stretch into.

7.6 Choosing line and page breaks by hand

When T_EX is in horizontal mode (inside a paragraph) and encounters a `\break`, it starts a new line. But it doesn't start a new paragraph: the next line is not indented, and the line just ended doesn't get the end-of-paragraph glue `\parfillskip`. As a result, the text must end at the right margin, and the glue generally gets overstretched. For this reason it is usually better to say `\hfill\break` instead of just `\break`.

Another thing to watch out for is that `...xyz\break` and `...xyz \break` don't have the same effect. In the second case, the blank following `xyz` will appear on the page, at the end of the line.

When T_EX is in vertical mode (between paragraphs) and sees `\break`, it starts a new page, same as if it had seen `\eject`. (But there is an important difference between the two: `\eject` will end the paragraph and break the page even if encountered in horizontal mode.) In this case, too, you should use `\vfill\break` rather than just `\break` if the spacing between paragraphs gets stretched unduly.

7.7 Floats

Suppose you want to save two inches of vertical spacing for a figure. If you type

```
\vskip 2in\centerline{Figure 5}
```

and there happens to be only one inch left at the bottom of the page, you'll find that the Figure 5 appears at the top of the next page, without any spacing before it (why?); also, the layout of the current page will look awful, since the spacing between paragraphs has to stretch to fill up the last inch of the page. If you use `\vglue` instead of `\vskip`, the required two inches will be saved at the top of the next page, but again the layout of the current page will be wrong.

Here's the right solution:

```
\midinsert\vglue 5cm\centerline{Figure 5}\endinsert
```

When T_EX encounters the pair `\midinsert...\endinsert`, it typesets the material between the two commands and stashes it away in a box. If there is room for this material on the current page, that is, if the height of the box is less than the amount of space left on the page, T_EX unboxes the material right there and moves on. But if there isn't enough room, T_EX saves the material for the top of the next page, and continues on the current page with the text that follows `\endinsert`. The migrating material constitutes a float.

Notice that even in a float you must use `\vglue` or `\null\vskip`, rather than `\vskip`, to leave space at the top; otherwise the `\vskip` would find itself at the top of the next page, and would consequently be discarded.

There is a variant for `\midinsert`: the pair `\topinsert...\endinsert` makes the intervening text migrate to the top of either the current page or the next page, depending on whether or not there is room on the current page.

7.8 A complete example

Here is the complete source of some lecture notes used for a math course at the University of Strasbourg, together with the corresponding T_EX output. You will recognize many of the layout hints mentioned in this chapter. All the commands used here are documented in the Dictionary.

```
\global\evenpagehead={\line{{\tenbf\folio}\quad
  \tenrm the\runningauthor\hfill}}
\global\oddpagehead={\line{\hfill\tenrm
  \the\runningtitle\quad\tenbf\folio}}
\runningauthor={N. Ikabruob}
\runningtitle={Linear algebra over $\Z$}

\def\eps{\varepsilon}
\def\Z{{\bf Z}} \def\R{{\bf R}} \def\Q{{\bf Q}}
\def\qed{\vbox{\hrule\hbox{\vrule\kern3pt
  \vbox{\kern6pt}\kern3pt\vrule}\hrule}}

{\parindent=0pt\obeylines
Universit\'e Louis Pasteur \hfill March 28, 1989
UFR de Math\'ematiques et d'Informatique
7, rue Ren\'e Descartes
67000 Strasbourg, France}

\vfill  % center title in available space
\centerline{\bf Linear Algebra over $\Z$}
\smallskip
\centerline{N. Ikabruob}
\vfill\eject
```

Université Louis Pasteur   March 28, 1989
UFR de Mathématiques et d'Informatique
7, rue René Descartes
67000 Strasbourg, France

Linear Algebra over Z

N. Ikabruob

```
\noindent
{\bf 1. Introduction}

\smallskip\noindent
In classical linear algebra, one proves the following results:

\smallskip
\item{(a)} every vector subspace of
$\R^n$ has a finite number of generators;

\item{(b)} every subspace of $\R^n$ has a basis, and two
bases have the same number of elements (its
{\it dimension\/});

\item{(c)} if $\{e_1,\ldots,e_k\}$ is a set of generators
for a subspace $M$, there is a subset of $\{e_1,\ldots,e_k\}$
that forms a basis for $M$;

\item{(d)} every set of linearly independent vectors
can be completed into a basis of $\R^n$;

\item{(e)} one can pass from any of the following
representations of a subspace to any other:

\smallskip
\itemitem{---} representation by generators (or by a basis),
\itemitem{---} representation by a system of equations,
\itemitem{---} parametric representation.
```

---

2   N. Ikabruob

1. Introduction

In classical linear algebra, one proves the following results:

(a) every vector subspace of $\mathbf{R}^n$ has a finite number of generators;

(b) every subspace of $\mathbf{R}^n$ has a basis, and two bases have the same number of elements (its *dimension*);

(c) if $\{e_1, \ldots, e_k\}$ is a set of generators for a subspace M, there is a subset of $\{e_1, \ldots, e_k\}$ that forms a basis for M;

(d) every set of linearly independent vectors can be completed into a basis of $\mathbf{R}^n$;

(e) one can pass from any of the following representations of a subspace to any other:

    — representation by generators (or by a basis),
    — representation by a system of equations,
    — parametric representation.

```
\smallskip
We will now examine the following problem: what happens
to these results when $\R^n$ is replaced by $\Z^n$?

\medskip\noindent
{\bf 2. Generalities}

\smallskip\noindent
Before we can answer this question in more detail, we
should familiarize ourselves with it and introduce certain
notions that will be useful later.

{\it Generators:\/} we say that the set $\{x_1,\ldots,x_r\}$
generates the subgroup $M$ of $\Z^n$ if every element of
$M$ is a linear combination of $x_1,\ldots,x_r$, with
integer coefficients (positive or negative).

{\it Linear independence over $\Z$:\/} we say that the vectors
$x_1,\ldots,x_r$ are linearly independent over $\Z$ if the
equality $a_1x_1+\cdots+a_kx_k=0$, with $a_i\in\Z$, forces
$a_1=\cdots=a_k=0$.

We notice right away that a set of vectors is linearly
independent over $\Z$ if and only if it is over $\Q$.
(Clear denominators!)  This lets us talk about linear
independence without specifying over what ring.

{\it Bases:\/} we say that $\{\eps_1,\ldots,\eps_r\}$ is a
```

Linear algebra over **Z**   3

We will now examine the following problem: what happens to these results when $\mathbf{R}^n$ is replaced by $\mathbf{Z}^n$?

2. Generalities

Before we can answer this question in more detail, we should familiarize ourselves with it and introduce certain notions that will be useful later.

Generators: we say that the set $\{x_1, \ldots, x_r\}$ generates the subgroup M of $\mathbf{Z}^n$ if every element of M is a linear combination of $x_1, \ldots, x_r$, with integer coefficients (positive or negative).

Linear independence over **Z***:* we say that the vectors $x_1, \ldots, x_r$ are linearly independent over **Z** if the equality $a_1 x_1 + \cdots + a_k x_k = 0$, with $a_i \in \mathbf{Z}$, forces $a_1 = \cdots = a_k = 0$.

We notice right away that a set of vectors is linearly independent over **Z** if and only if it is over **Q**. (Clear denominators!) This lets us talk about linear independence without specifying over what ring.

Bases: we say that $\{\varepsilon_1, \ldots, \varepsilon_r\}$ is a basis for a subgroup M of $\mathbf{Z}^n$ if it generates M and $\varepsilon_1, \ldots, \varepsilon_r$ are linearly independent. For example, the canonical basis is a basis for $\mathbf{Z}^n$.

```
basis for a subgroup $M$ of $\Z^n$ if it generates $M$ and
$\eps_1,\ldots,\eps_r$ are linearly independent.  For
example, the canonical basis is a basis for $\Z^n$.

{\it Remark:\/} If $\{\eps_1,\ldots,\eps_r\}$ is a basis for
a subgroup $M$ of $\Z^n$, it is also a basis for the vector
subspace of $\Q^n$ generated by $M$.  Thus two bases over
$\Z$ have the same number of elements.  We'll have a notion
of dimension as soon as we establish the existence of one
basis\dots

{\it A counterexample:\/} Items (c) and (d) in section 1 are
no longer true.  For consider the subgroup $M$ of $\Z^2$
generated by the vectors
$$
\eps_1=(2,0),\quad\eps_2=(1,3),\quad\eps_3=(0,9).
$$
These vectors are not linearly independent.  Any two of
the three are linearly independent, but then they don't
generate $M$!  Notice also that $M\neq\Z^2$.

{\it Unimodular matrices:\/} A unimodular matrix is an
invertible matrix $A\in M(n,\Z)$ whose inverse also has
integer coefficients.

\proclaim Theorem. If $A$ be an invertible
square matrix with integer coefficients, $A^{-1}$ has
integer coefficients if and only if $\det A=\pm 1$.
```

---

4 N. Ikabruob

Remark: If $\{\varepsilon_1, \ldots, \varepsilon_r\}$ is a basis for a subgroup M of $\mathbf{Z}^n$, it is also a basis for the vector subspace of $\mathbf{Q}^n$ generated by M. Thus two bases over $\mathbf{Z}$ have the same number of elements. We'll have a notion of dimension as soon as we establish the existence of one basis...

A counterexample: Items (c) and (d) in section 1 are no longer true. For consider the subgroup M of $\mathbf{Z}^2$ generated by the vectors

$$\varepsilon_1 = (2,0), \quad \varepsilon_2 = (1,3), \quad \varepsilon_3 = (0,9).$$

These vectors are not linearly independent. Any two of the three are linearly independent, but then they don't generate M! Notice also that $M \neq \mathbf{Z}^2$.

Unimodular matrices: A unimodular matrix is an invertible matrix $A \in M(n, \mathbf{Z})$ whose inverse also has integer coefficients.

Theorem. *If A be an invertible square matrix with integer coefficients, A^{-1} has integer coefficients if and only if $\det A = \pm 1$.*

```
\smallskip\noindent
{\it Proof:\/} we have $\det A \det A^{-1}=1$. If $A$ and
$A^{-1}$ have integer coefficients, their determinant is
also an integer.  Thus $\det A=\pm 1$. Conversely, if
$\det A=\pm 1$, we deduce that $A^{-1}$ also has integer
coefficients, because it can be expressed as an integer
multiple of the matrix of cofactors of $A$.%
\enspace\qed

\smallskip
Because of the preceding theorem, unimodular matrices
form a {\it group\/} under matrix multiplication.  We denote
this group by ${\rm GL}(n,\Z)$.

\bye
```

Linear algebra over **Z** 5

Proof: we have $\det A \det A^{-1} = 1$. If A and A^{-1} have integer coefficients, their determinant is also an integer. Thus $\det A = \pm 1$. Conversely, if $\det A = \pm 1$, we deduce that A^{-1} also has integer coefficients, because it can be expressed as an integer multiple of the matrix of cofactors of A. $\square$

Because of the preceding theorem, unimodular matrices form a *group* under matrix multiplication. We denote this group by $\mathrm{GL}(n, \mathbf{Z})$.

7.9 Penalties: or, the carrot and the stick

This section is a bit more technical (but not hard to understand). Skip it the first time around, and consult it when you have problems with page layout and page breaks.

TEX decides on lines and pages breaks after considering many different possibilities. It chooses among them on the basis of the accumulated demerits for the various elements of each configuration. Demerits are a measure of the ugliness of a configuration, and they can come from many sources: lines that are stretched or compressed too much, excessive hyphenation, club or widow lines (first and last lines of a paragraph stranded on a page by themselves), and so on. You can influence the computation of demerits, and consequently the outcome of the line- and page-breaking process, by using the `\penalty` command.

A penalty is a number between -10000 and 10000 that you place at any spot where you want to encourage or discourage a break. In computing the demerits for any configuration that includes a break at that point, TEX will take the penalty into account, increasing the demerits if the penalty is positive and decreasing it if the penalty is negative. The rules for how the penalty is taken into consideration are complex, but here's what you have to know to get going:

• Basically, TEX will consider line or page breaks only at glue (between words, between paragraphs, or caused by `\hskip`, `\vskip` and the like), or at a penalty.

• A penalty is incorporated to the demerits of a break that *follows* it. Penalties don't apply retroactively: you must place them before any glue if they are to work.

• A penalty of 10000 is so high that it prevents a break altogether; a penalty of −10000 is so highly negative—that is, it indicates such a good breakpoint—that TEX will always break there.

To discourage a break at some space or glue, you can say `\penalty 100` or `\penalty 200` just before it; this will make that break that much less attractive in TEX's eyes. If instead you say `\penalty -100`, you encourage a break at that spot. As an example, the definition of an unbreakable space ˜ says `\penalty 10000\ `. If TEX tries to break at such a space, it has to take into account the penalty of 10000, which is just too high.

Here are the main commands that plain TEX provides to help with page layout. Many of them are simply shorthands for some penalty or another, and so can be used both in horizontal mode (for line breaks) and in vertical mode (for page breaks):

• `\allowbreak` stands for `\penalty 0`. It normally has no effect next to glue, since a break is allowed there anyway; but it can be very useful at places where TEX would not consider a break otherwise: for example, within certain math formulas.

• `\nobreak` stands for `\penalty 10000`. It completely forbids a line or page break.

• `\break` stands for `\penalty -10000`. It forces a line or page break.

Because the action of these commands depends on the mode, they sometimes have unexpected consequences. A common mistake is to say

```
... in vertical mode (for page breaks):
\nobreak
\smallskip
\meti{$\bullet$} ...
```

in order to guarantee that the list starts on the same page as the preceding paragraph. This doesn't work because the `\nobreak` is read in horizontal mode, so it prevents a *line* break, not a page break! (What's the solution?)

Other commands first put TEX in vertical mode, then insert the penalty:

• `\eject` stands for `\par\penalty -10000`. It finishes off the current paragraph and forces a line or page break.

• `\supereject` stands for `\par\penalty -20000`, a value not used otherwise. It not only breaks the page, but also forces any floats, footnotes, etc. that may be in memory to be printed before TEX goes any further. Useful at the end of a chapter. Plain TEX incorporates `\vfill\supereject` into the `\bye` macro, the recommended way to finish a run.

- \goodbreak stands for \par\penalty -500. It finishes the current paragraph and hints that this is a good place for a page break (but only if the page is pretty much complete already).

Finally, there are some commands that combine (negative) penalties with vertical spacing. They are very useful in practice:

- \filbreak stands for \par\vfil\penalty -200\vfilneg. It says this is a good place for a page break, even if the page is not complete. The spring \vfil will fill up the rest of the page if the break is chosen. If not, it will be canceled by the \vfilneg, and it will as if nothing had happened. (This macro should be used carefully. If you use it after each paragraph, TeX won't break paragraphs; it will instead leave white space at the bottom of each page where it can't fit a whole paragraph.)

- \bigbreak combines a \penalty -200 and a conditional \vskip 12pt plus 4pt minus 4pt. That is, the skip is not put in if the \bigbreak command was immediately preceded by a skip of 12 pt or more. Also, if the \bigbreak was preceded by a skip of less than 12 pt, that skip is canceled. In particular, two consecutive \bigbreak s have the same effect as one.

- \medbreak and \smallbreak work just like \bigbreak, but the penalties and skip amounts are halved and quartered, respectively.

Naturally, you can use \penalty directly, with any value you feel like. But using carrots and sticks is a subtle art that one learns gradually, and you may be mystified at first by the results of your experiments. Here are some hints:

- Remember that a penalty has no effect on glue (either horizontal or vertical) that comes before it. In particular, a \nobreak won't help if it is preceded by glue. (Sometimes you can't figure out where the glue comes from: it may have been put there by some macro. Try to change TeX's mind by inserting a negative penalty a little above or below.)

- Don't be heavy-handed. Penalties enter into the computation of demerits after being squared, so a \penalty 500 goes a long way. Using very high penalties all over the place will just lead to unpredictable results, because it will upset the "balance of power" set up by plain TeX.

8
Boxes

8.1 What is a box?

To TEX, everything is a box! TEX has no idea what an 'A' or an integral sign looks like. It thinks of them simply as boxes with certain dimensions. When characters are put together to form a line, the line itself becomes a box, and these boxes assemble in even bigger boxes, and so on. We could say that TEX's job consists of two things: creating boxes and putting them together.

Boxes can be implicit or explicit. Implicit boxes are the most common—every character and every line of text is one. Explicit boxes are created by the commands \hbox, \vbox, \vtop and \vcenter, which we will study in this chapter.

Once TEX has created a box, it is no longer interested in its contents, at least temporarily. The box becomes an outline, an imaginary rectangle whose dimensions are its only concern. Here is what TEX sees in a box:

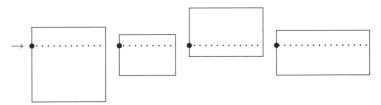

The dotted line inside each box is called its baseline; its importance is that boxes arranged side by side are aligned according to their baselines, as in the figure. You can think of the baseline as the hole through a bead, and of the several boxes shown as a string of beads with a thread going through all of them. The reference point of a box is the left endpoint of its baseline; the reference point of the leftmost box in the picture is indicated by an arrow.

The dimensions of a box are its width, its height (the elevation above the baseline) and its depth (below the baseline). The leftmost box in the previous figure has a width of .8 in, a height of .2 in and a depth of .6 in. Each dimension of a box is normally positive, but it can be zero or even negative: for instance, the macros \rlap and \llap (section 5.10) put their argument into a box of width zero.

8.2 Putting boxes together

We saw in chapter 3 that the fundamental difference between vertical and horizontal mode is how boxes are put together: on top of one another, like a stack of pancakes, or side by side, like a string of beads. We now investigate this difference further.

Stacking boxes up

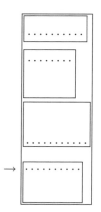

When TEX reads \vbox, \vtop or \vcenter, followed by a left brace, it switches to internal vertical mode: it starts stacking one above the other the boxes encountered or created from then on, aligning their reference points vertically. It continues to do this until it reads the matching right brace. The result is again a box, called a vertical box. For example,

\vbox{\box1\box2\box3\box4}

gives the big box on the left, where \box1, \box2, \box3 and \box4 refer to the small boxes inside; \box1 is at the top.

The width of the outer box is the maximum width of the component boxes (for clarity, we drew the outer box slightly larger than it really is). TEX automatically puts in a bit of glue between component boxes; we'll discuss this in more detail later.

TEX can also be in vertical mode without a surrounding box: in fact, that's the state it starts in. In this so-called ordinary vertical mode TEX is building up the current page, as if it were a big vertical box; but, unlike a vertical box, a page has a predetermined height, and when that height is reached, TEX ships out the page and starts a new one. In internal vertical mode, by contrast, material just keeps piling up, until the vertical box is finished.

Why are there three commands to make vertical boxes? They behave identically, except at the very end. The baseline of a box obtained with \vbox coincides with the baseline of the last, or lowermost, component box. The baseline of a \vtop, on the other hand, coincides with the baseline of the top box in the stack. Finally, with a \vcenter—which is allowed in math mode only—the resulting box has its bead hole right in the middle. A figure will help make the difference clearer:

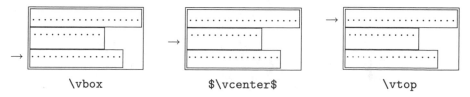

\vbox $\vcenter$ \vtop

Stringing boxes together

When T_EX encounters \hbox, it switches to restricted horizontal mode. This means it strings together side by side the material that follows within braces; the result is a horizontal box, exactly big enough to fit all the material. The component boxes in an \hbox are aligned by their baselines:

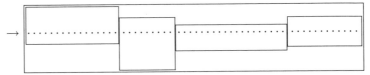

\hbox{box1\box2\box3\box4}

(Here again, the resulting box is shown slightly bigger than it actually is.) The width of the big box is the sum of the widths of the constituent boxes, because, unlike the vertical case, no glue is put between them. The height and depth of the big box are the maximum height and depth of the boxes inside.

When T_EX is composing a paragraph, it is likewise in horizontal mode, called ordinary. The difference between ordinary and restricted horizonal mode is that in the former T_EX will break up the resulting box into chunks of length \hsize — lines spanning the width of the page—and will always create at least one such chunk. In an \hbox, on the other hand, exactly one line is created, and it can be of any length, depending on the material inside.

8.3 What goes in a box?

In a word, everything. Anything that goes on the page is put into a box at some point. But different types of material obey different rules, so we must consider them separately. The basic types are five:

Characters

Characters are boxes in the sense that they have width, height and depth, but they are peculiar in some respects. For one thing, they can only occur in horizontal mode: if T_EX is in vertical mode when it sees a character, it immediately goes into ordinary horizontal mode and starts a paragraph, which is later cut up into lines of width \hsize (see chapter 6 and the end of the previous section). Remember, then: when you are in vertical mode, a single character triggers the creation of a whole line!

Whether in ordinary or in restricted horizontal mode (caused by an \hbox), characters are strung together side by side, aligned according to their baselines:

In this figure, each character is shown surrounded by the box it defines; the ligature 'fi' counts as one character. (As usual, boxes are shown slightly bigger than they actually are.) What TEX actually sees is something like this:

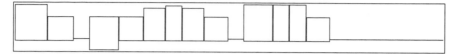

Another way in which characters are special is that they are sometimes separated or brought closer together by automatic kerns. This was ignored in the figure above.

Glue

Glue is either horizontal or vertical. Horizontal glue is for horizontal and math modes only; if TEX is in vertical mode and sees \hskip it switches to ordinary horizontal mode, same as if it sees a character. Vertical glue is for vertical mode only; if TEX sees \vskip while in ordinary horizontal mode it finishes the current paragraph, whereas in restricted horizontal mode it gives an error message and tries to finish the surrounding \hbox .

On the other hand, \kern doesn't change the mode; it simply is interpreted differently, as a horizontal or vertical kern, according with the mode.

Unlike the other types of material, glue has only one dimension: horizontal glue has width, and vertical glue has height. This means that glue inside a horizontal box doesn't affect the box's height or depth, and glue inside a vertical box doesn't affect the box's width.

Rules

Rules are horizontal or vertical straight lines; we'll discuss them in more detail in section 8.10. There are horizontal and vertical rules, to be used in vertical and horizonal mode, respectively. Thus, TEX goes into vertical mode, if it is not there already, when it sees \hrule ; any following boxes will stack up. Conversely, a \vrule puts TEX in horizontal mode, and boxes following it are strung together horizontally.

An \hrule is placed immediately next to the preceding and following boxes, without any vertical glue being added.

Explicit boxes

Although there are also vertical and horizontal boxes, they are much more liberal in their associations than either glue or rules: either type can occur in either mode. Inside the box, of course, the mode changes accordingly, but once the box is finished, the mode reverts to what it was just before the box was read (see also section 3.3). So boxes, of either type, get piled up when they occur inside a \vbox , but placed side by side when they occur inside an \hbox .

Mathematical formulas

Math formulas are built in a special way from material that is surrounded by dollar signs $. They are like characters in that they should only occur in horizontal mode;

TEX will start a new paragraph if it sees a `$` while in vertical mode. Material surrounded by double dollar signs `$$` is even more special: since it is meant to be displayed on a line by itself, interrupting the current paragraph, it can't exist inside an `\hbox`, but only in ordinary horizontal mode. TEX will basically ignore a `$$` inside an `\hbox`.

Some examples

Here are some experiments to flesh out this theory a little. We start with a `\vtop`:

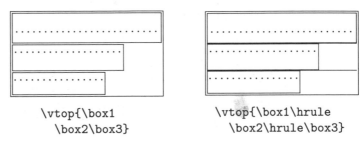

```
\vtop{\box1            \vtop{\box1\hrule
\box2\box3}            \box2\hrule\box3}
```

In the box on the left there are three stacked boxes. In the box on the right, we put in an `\hrule` between each pair of boxes; this doesn't perturb the vertical mode, so everything is still stacked up. Notice, by the way, that the interbox space disappeared when we put in the rules. The width of the resulting box is always the maximum of the widths of the components.

Let's throw in some text now:

```
\vtop{\box1 Once upon a time \box2\box3}
```

It's all head over heels! To understand what's going on, let's follow TEX's reasoning step by step. When it starts the `\vtop`, it is in vertical mode. It sees `\box1`, gets it from its memory, and places it as the topmost box in the stack that it's building. Next it sees the character 'O'. This makes it go into horizontal mode! So a paragraph starts, which has no reason to end until TEX sees the right brace. Then the paragraph must end, because the enclosing `\vtop` ends. So the paragraph has three words and two boxes, `\box2` and `\box3`, which all fit on one line:

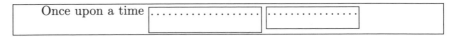

(notice the indentation at the beginning of the line). This one-line paragraph is now placed underneath `\box1`, aligned on the left.

To check your understanding, try to explain the outcome of the next two experiments. To help you out, the box caused by the text is drawn in both cases. On the left this box would occupy the whole width of the page, so we made the width of the "page" (the \hsize) small. Notice that the text is indented in the first case, but not in the second: there is no indentation inside an \hbox .

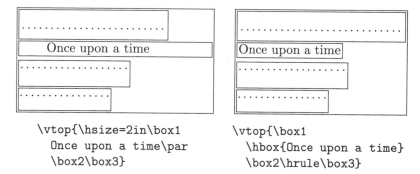

```
\vtop{\hsize=2in\box1
    Once upon a time\par
    \box2\box3}
```
```
\vtop{\box1
    \hbox{Once upon a time}
    \box2\hrule\box3}
```

So far, we would have gotten the same result by saying \vbox or \vcenter instead of \vtop . The only difference between the three types of vertical boxes is the placement of the baseline, not the internal organization.

We conclude with a roller coaster of \hbox es:

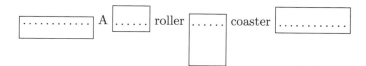

We typed \ \box1\ A \box2\ roller \box3\ coaster \box4 to get this; notice that we had to go into horizontal mode "by hand" with a \ .

8.4 Creating a box: summary

To summarize, you create an explicit box when you type \hbox , \vbox , \vtop or \vcenter , followed by material in braces. The material can be anything compatible with the mode of the box. As usual, the braces delimit a group, so if you say \hbox{\bf ...} the font change will only apply inside the box.

In the beginning, you may be perplexed with so many types of boxes. But choosing among them will soon become second nature to you. What you have too keep in mind is that:

• An \hbox always contains exactly one line. The line can grow to any length, according to what's inside, but there is no point in making it any longer than the page width \hsize . One often uses an \hbox to keep a few words from breaking across lines (see the next section), or to arrange several boxes side by side.

• A vertical box can contain anything. If it contains any horizontal mode material (characters, horizontal glue, \vrule s, math) that is not "protected" inside an \hbox , this material forms whole paragraphs, so the vertical box has width \hsize . If you want to typeset text inside a vertical box, then, you should each break the text into lines that go in individual \hbox es, or set the \hsize inside the vertical to box to the desired box width.

- Generally speaking, the baseline of a \vbox is near the bottom, so the box "sticks up" when placed next to others. The baseline of a \vtop is near the top, and that of a \vcenter is right at the middle. A \vcenter can only be used in math mode.

8.5 Storing a box

The description of a box can be very complicated, and it can be confusing to try to build it in the middle of the text where it is to appear. It's much better to build it and *store* it for later use:

```
\setbox1=\hbox{...}
\setbox17=\vbox{...}
\setbox48=\vtop{...}
```

There are 256 slots in TEX's memory for boxes, and they're called \box0, \box1, ..., \box255. The command \setbox2 tells TEX to store the box that follows in slot 2. When that box is created, then, it doesn't appear on the output; it only gets used when you say \box2. A \setbox assignment, like any other, has effect only until the end of the group in which it is made, unless it is preceded by \global. The = in the assignment is optional.

Some of the box slots are appropriated for certain uses, and you shouldn't mess with them: for example, box 255 contains the output page. In fact, you should probably not use box numbers at all. Plain TEX lets you "reserve" and give a name to a box that is otherwise unused:

```
\newbox\toto
\setbox\toto=\hbox{...}
...
\box\toto
```

The first command, \newbox\toto, needs to be given only once; from then on you can use the name \toto as if it were a number, to refer to this particular box, as many times as you want (compare with \newfam in section 4.7).

Boxes 1 through 9 are saved for scratch use, so it's OK to say \setbox1=... for very short-term storage. Between the time you set the box and the time you use it there should be no intervening commands that might reset the box.

The only way to store a centered box (\vcenter) is to make a math formula out of it, and put the expression in an \hbox , like this:

```
\setbox\toto=\hbox{$\vcenter{...}$}
```

Don't forget the dollar signs. Also, don't try to put the $\vcenter...$ directly into a \vbox ; since math formulas are only legal inside a paragraph, TEX will make a paragraph out of it, and the resulting box will have width \hsize .

Using a box

You now know how to store a box and how to use it with the `\box` command. But there are other commands for using the contents of a box:

`\box`, `\copy`, `\unhbox`, `\unvbox`, `\unhcopy`, `\unvcopy`.

When you say `\box3`, TEX puts the contents of box 3 at the current point in the text. But it also erases box 3 altogether, and its contents are lost. If you want to use the contents of box 3 without erasing it, you must say `\copy3`. The same is true if you're dealing with a box by name: say `\copy\toto`, rather than `\box\toto`.

The reason `\box` behaves in this way is economy of space: since boxes generally contain a lot of stuff, and most often are used only once, TEX tries to free up memory by making the read-once behavior the default.

TEX won't split the contents of a box. If you say

```
\setbox\Max=\hbox{``I'LL EAT YOU UP!''}
\parindent=0pt

The night that Max wore his wolf suit and made
mischief of one kind and another his mother called
him ``WILD THING!'' and Max said \box\Max
```

you get an overfull line, because TEX won't go inside box `\Max` to break between the two words there:

The night that Max wore his wolf suit and made mischief of one kind and another his mother called him "WILD THING!" and Max said "I'LL EAT YOU UP!"■

Compare with what happens when you use the box by saying instead `...and Max said \unhbox\Max`:

The night that Max wore his wolf suit and made mischief of one kind and another his mother called him "WILD THING!" and Max said "I'LL EAT YOU UP!"

Here TEX effectively unboxed the contents of the box before inserting them into the paragraph. The result is that a break can be made between the two words, as if they had never been boxed in the first place. TEX even lets the glue between the elements of an unboxed box stretch and shrink, which is not the case inside a box that has already been wrapped up.

But you shouldn't expect to be able to modify the unboxed contents in any way: once they've been set, TEX won't go back! For example, suppose the definition of the box `\Max` is preceded by a change of font:

```
\bf\setbox\Max=\hbox{``I'LL EAT YOU UP!''}\rm
```

When you `\unhbox` the box, here's what you get:

The night that Max wore his wolf suit and made mischief of one kind and another his mother called him "WILD THING!" and Max said **"I'LL EAT YOU UP!"**

This behavior at first appears contradictory: why does the spacing adjust itself as if "I'LL EAT YOU UP" were a part of the paragraph, but the font is different? It's because TₑX is splicing in the *contents* of the box, not reading its definition again. Things would be different if you had said `\def\Max{''I'LL...}`; then whenever you used the macro `\Max` the phrase would appear in the current font.

The command `\unvbox` works just like `\unhbox`, but is used for vertical boxes, those created with `\vbox` or `\vtop`. Like `\box` itself, `\unhbox` and `\unvbox` erase the contents of a box when they unbox them. The peculiarly named commands `\unhcopy` and `\unvcopy` unbox a box without erasing its contents.

To make a long story short, use `\box` when you need to keep the contents together, and will use them only once. Use `\unhbox` and `\unvbox` when you need more flexibility. And use `\copy`, `\unhcopy` or `\unvcopy` when the contents are needed more than once.

8.6 The baseline

The baseline of a box is determined by its contents. We have seen that:

- the baseline of a `\vtop` coincides with the baseline of its first component;
- the baseline of a `\vbox` coincides with that of its last component;
- the baseline of a `\vcenter` goes through the middle; and
- the baseline of an `\hbox` is the common baseline of all its sub-boxes.

But a box can contain things other than boxes. How is the baseline determined then? Characters and rules are no problem: they have baselines too, and behave just like boxes for the purpose of building up other boxes (section 8.3). Glue affects the positioning of the baseline as follows: a `\vbox` that ends with glue has its baseline all the way at the bottom; a `\vtop` that begins with glue has its baseline all the way at the top. For the other two types the rule doesn't change.

Changing the baseline: a drastic remedy

Suppose you're given a box made with `\vbox`, say box 1, and need to move its baseline to the top, as if it were a `\vtop`. Saying `\vtop{\box1}` won't work, because the `\vtop`, having only one box inside, must inherit its baseline. This is shown on the left:

Paris was under siege, starving and at her last gasp. The sparrows were disappearing from the roofs, and the city's sewers were being depopulated. People were eating anything they could find.

→ One bright morning in January Monsieur Morissot, a watchmaker by trade but an idler by necessity, was walking sadly along the outer boulevard with an empty stomach. . .

→ . . . and his hands in the pockets of his uniform trousers when he came face to face with a comerade in arms whom he recognized as an old friend. It was Monsieur Savage, a riverside acquaintance.[1]

```
\setbox1=\vbox{...}
\vtop{\box1}
```

```
\setbox1=\vbox{...}
\vtop{\kern0pt\box1}
```

```
\setbox1=\vbox{...}
\vtop{\unvbox1}
```

---

[1] Guy de Maupassant, *Two Friends*, translated by Roger Colet.

One solution, based on the rules stated before, is to tack some glue onto the box before wrapping it inside `\vtop`. Naturally, the glue shouldn't take up any space:

`\vtop{\kern 0pt\box1}`

This gives the middle part of the triptych. But notice that the baseline is now all the way at the top, rather than coinciding with the baseline of the first line inside, as would have been the case if the paragraph had been set in a `\vtop` in the first place. The individual lines are still not available to the outer box.

There is a better solution, at least as long as the paragraph isn't packed too deeply inside other boxes. The idea is to first unbox the contents of the `\vbox`, freeing up the individual lines; TeX will then be able to use the baseline of the first line as the baseline of the outer box. This is shown on the right.

Similarly, to move the baseline of a `\vtop` to the bottom, you can put it in a `\vbox` followed by glue, or you can unbox its contents inside the `\vbox`. Moving the baseline to the center is easier: you always say `\hbox{$\vcenter{\box1}$}`, no matter how box 1 was created.

Changing the baseline: fine-tuning

In spite of all precautions, it can happen that boxes aligned by their baselines don't look quite right, and one wants to move them up or down individually. The commands `\raise` and `\lower` let you do that; they only work in horizontal mode, that is, while TeX is laying boxes side by side. For example, when you say

`\raise 5pt\box1`

TeX raises `\box1` by 5 pt before adding it to the paragraph or `\hbox` that it is setting. It's as if the baseline of `\box1` had been lowered by 5pt; and in fact if you say `\setbox1=\hbox{\raise 5pt\box1}`, the net result is that the baseline of `\box1` is now 5 pt lower. The height of `\box1` increases by 5 pt, and its depth decreases by the same amount.

The `\raise` command must be followed by a dimension and then a box of any type, made on the spot or retrieved from one of the slots. Nothing else will work! For example, `\raise 5pt{\vbox{...}}` gives an error; the braces around the `\vbox` are wrong.

In the figure below, boxes 1, 2 and 3 have height 7, 5 and 9 mm, respectively. We raise each one so that its top is 10 mm above the baseline of the enclosing box, which is marked by an arrow (as usual, the enclosing box is shown slightly bigger than it is in actuality):

`\hbox{\raise 3mm\box1\raise 5mm\box2\raise 1mm\box3}`

Naturally, \lower is the opposite of \raise. It is used, for example, in the definition of the TEX logo in section 5.11. You can get by without it, if you want, because \raise -.1in\box5 has the same effect as \lower .1in\box5.

Moving a box horizontally

As we have seen, \raise and \lower work only in horizontal mode. They have a vertical mode couterpart: you can shift a box to the left or to the right using \moveleft and \moveright, followed by a dimension and a box: \moveleft .2in\box2. (You might be tempted to achieve the same thing with \hskip -.2in\box2; but this will start a paragraph, while \moveleft leaves you in vertical mode.)

8.7 The dimensions of a box

Examining the dimensions of a box

You can refer to the dimensions of a box stored in one of the memory slots by saying \ht, \dp and \wd, followed by the box number or name. To get the height of box 1 to appear on your screen, put the command \showthe\ht1 in your file; TEX will stop at that point, display the number you want, and wait for a carriage return to proceed.

If your box has not been stored there is no way to refer to its dimensions; you must assign it a slot first.

How the dimensions are determined

By default, the dimensions of a box are fixed by its contents, as we saw in the preceding sections. To recap:

• For a horizontal box, the height and depth are the maximum height and depth of the boxes inside. The width is the sum of the widths of the boxes inside, plus glue, if any.

• For a vertical box, the width is the maximum width of the boxes inside. The total vertical dimension (sum of height and depth) is the sum of the vertical dimensions of the boxes inside, plus glue, if any. This vertical dimension is split evenly between height and depth for a \vcenter; it goes mostly toward the height for a \vbox; and goes mostly toward the depth if the box is a \vtop.

Once again, if there is even a single character "out in the open" in a vertical box, TEX will start a paragraph, with lines of length \hsize. This means the box will have width at least \hsize. When you put text directly in a vertical box, then, don't forget to fix the \hsize accordingly:

$$\text{\textbackslash vbox\{\textbackslash hsize=2.5in...\}}$$

Dimensions fixed from the outside

Depending on the type of box, you can constrain one or another dimension to have a fixed value. If you say \vbox to 2in{...} or \hbox to 4cm{...} you will get, respectively, a \vbox of height 2 in and an \hbox of width 4 cm. (You can also say `\vtop to...` and `\vcenter to...`, but these constructions are best avoided, since they don't do what you expect. If you really want to know what they do, see pages 81, 290 and 443 of *The TEXbook*.)

To satisfy the constraint, TEX will stretch or shrink any glue that might be present in the box. If doing that requires going beyond the available elasticity, you get an overfull or underfull box, accompanied by an error message. For this reason, it's generally a good idea to use springs, which have infinite stretchability. After you've had some experience you may want to try the springs with infinity shrinkability, `\vss` and `\hss`, for special effects (cf. the definition of `\centerline`).

As a special case, `\vbox to 10mm{}` creates an empty box 10 mm tall, and having height and width zero. No springs are necessary!

To use the contents of a stored box inside a box whose dimension is predetermined, `\unvbox` and `\unhbox` often come in handy:

```
\setbox1=\hbox{Once upon a time}
$|$\copy1$|$ .................................... |Once upon a time|
$|$\hbox to 1.2in{\copy1}$|$ ................. |Once upon a time  |
$|$\hbox to 1.2in{\unhbox1}$|$ .............. |Once upon a  time|
```

Using the `to` construction, you can make boxes with different contents conform to the same dimensions. For instance, to create a box with the same height as `\box5`, to place the two side by side in a display, say `\setbox6=\vbox to \ht5{...}`.

There is another command to set the dimension of a box from the outside:

```
\setbox1=\hbox{Gone with the wind}
$|$\copy1$|$ .................................... |Gone with the wind|
$|$\hbox spread 3pc{\copy1}$|$ ........ |Gone with the wind     |
$|$\hbox spread 3pc{\unhbox1}$|$ ..... |Gone   with   the   wind|
```

You've probably caught on already: `\hbox spread 3pc` *adds* 3 pc to the natural width that the box would have otherwise. (Notice that `spread`, like `to`, doesn't have a backslash.) Here again you should think of using `\unhcopy` or a spring, or TEX may complain that the box if overfull or underfull.

Naturally, you can also use `spread` with vertical boxes (of all types).

8.8 Some practical situations

We start with a macro to build a blank box with specified height, depth and width, represented in the definition by `#1`, `#2`, `#3`. It works "from the inside out"— first `\vbox to #1{\vfil}` makes a "box" of the right height and zero width and depth, then `\vtop spread #2{...\vfil}` stretches it down, and finally `\hbox spread #3{\hfil...}` stretches it sideways. The result of each command is passed to the next, surrounding, command:

```
\def\emptybox#1#2#3{\hbox spread #3{\hfil
  \vtop spread #2{
    \vbox spread #1{\vfil}
    \vfil}}}
```

Here is one of the most common horizontal arrangements of boxes placing boxes side by side(notice the alignment at the top):

It is obtained like this:

```
\setbox1=\vtop{...} \setbox2=\vtop{...}
\centerline{\box1\quad\vrule\quad\box2}
```

The material is stored into vertical boxes of the `\vtop` variety, which are then placed side by side in a horizontal box. To adjust the spacing we use horizontal springs. Notice how simple it is to place a vertical rule between the boxes.

If `\box1` and `\box2` are made up of other boxes, their tops may not coincide when the baselines align. This is sometimes what you want—when the first thing in the boxes is a line of text, the baselines should coincide, not the top of the lines. But if you really want the tops aligned, start each `\vtop` with a box of zero height, called `\null`. This will move the baseline of the big boxes to the top.

Many of the figures in this book are made up of two boxes set side by side. Most often they are aligned at the center:

The solution is similar: the boxes (of any type) being already built, we say

```
$$\line{$\hfill\vcenter{\box1}\quad \vcenter{\box2}\hfill$}$$
```

To place braces next to a box, center the box, then build the braces by surrounding it with `$\left\{...\right.$` (see section 11.13). The period after `\right` is part of the construction.

some words $\left\{$

```
\centerline{some words $\left\{\ \vcenter{...}\right.$}
```

Next, we place a legend underneath a memory box, say box 1.

```
\vbox{\hsize=\wd1
    \box1
    \medskip
    \centerline{Legend}
}
```

Legend

Changing the dimensions of a box

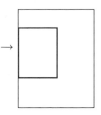

It is possible, and often useful, to change the dimensions of a memory box without changing its contents. In the figure on the left, the thin outline represents a memory box, say `\box1`, whose height, depth and width are $16, 20$ and 10 mm. By saying

$$\texttt{\textbackslash ht1=5mm \textbackslash dp1=8mm \textbackslash wd1=10mm}$$

all the dimensions are halved, and we get the smaller box indicated with a heavy outline. All the changes are with respect to the reference point.

This idea is useful if you have several memory boxes with legends, as we discussed above, and they must be placed side by side. Unless the boxes have the same depth, their legends won't align. You can make their depths the same if you know, for example, that box 1 is deeper than box 2:

```
\dp2=\dp1
\setbox1=\vtop{\hsize=\wd1
   \box1
   \medskip
   \centerline{Legend 1}}
\setbox2=\vtop{\hsize=\wd2
   \box2
   \medskip
   \centerline{Legend 2}}
\centerline{\box1\qquad\box2}
```

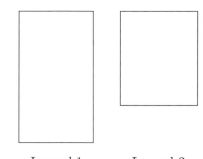

Legend 1   Legend 2

Another very useful application is in plain TeX's `\smash` macro. If you say `\smash{...}`, TeX will typeset the material in braces, but pretend that it has no height or depth! The baseline of the resulting box is the same as that of the material inside. The definition is basically very simple: it puts the material in a box, changes the height and depth of the box to zero, and writes it out:

```
\def\smash#1{\setbox0=\hbox{#1} \ht0=0pt \dp0=0pt \box0}
```

You will find all sorts of applications for this macro. Certain characters, for instance, create a bit of extra space between lines when used as superscripts.[†] The result is often unesthetic. To avoid this, you must prevent TeX from seeing the height of the superscript when it stacks up the line; the following code will do that:

```
...superscripts\footnote{\smash{$^{\dag}$}}
```

Here's another case in which `\smash` comes in handy:

$$R^{(1)}(a, u, v) = \sum_{|p - p_1| \leq N} \varphi(2^{-p}D)(S_{p - N_0} - S_{p_1 - N_0})(a)$$
$$\times \varphi(2^{-p_1}D)R_2(u, v)$$

[†] Like this.

To build up this formula, we stored each line in a box, then wrote

```
$$\displaylines{\qquad\smash{\box2}\hfill\cr
                \kern 65mm\box3\hfill\cr}$$
```

Without this precaution, the lines would be set too far apart, because the depth of the summation sign.

Corner letters, or dropped caps

M AN is but a reed, the most feeble thing in nature; but he is a thinking reed. The entire universe need not arm itself to crush him. A vapor, a drop of water suffices to kill him. But, if the universe were to crush him, man would still be more noble than that which killed him, because he knows that he dies and the advantage which the universe has over him; the universe knows nothing of this.[2]

This elegant effect was achieved with the `\cornerbox` macro, which employs many of the ideas we've discussed so far. This macro carves out the upper left corner of a paragraph, using `\hangindent` and `\hangafter`, and places there some other material. Here is its definition:

```
\def\cornerbox#1#2#3{\setbox1=\hbox{#1} \dp1=0pt
    \par\hangindent\wd1 \hangafter-#2 \noindent
    \hskip-\wd1 \raise#3 \box1 \ignorespaces}
```

The material that should go in the corner is the first argument to the macro, represented by `#1`. It is put inside box 1, whose depth is then declared to be zero, to prevent it from creating extra space between the first and second lines. Then `\hangindent\wd1` and `\hangafter-#2` carve out a corner with same width as `\box1` and a depth corresponding to `#2` lines; this value must be determined by hand. The paragraph proper starts with `\noindent`. Then `\hskip-\wd1` backtracks to the left margin, and we place the contents of `\box1` there, raised by an amount `#3` which can be fine-tuned. Finally, `\ignorespaces` eliminates unwanted blanks that may creep in right after the macro is called. Here's how `\cornerbox` was called to do the paragraph above:

```
\font\huge=cmr12 at 36pt
\cornerbox{\vtop{\kern 0pt\hbox{\huge M\kern 2pt}}}{3}{6pt}
AN is but a reed ...
```

Here we used `\cornerbox`. This time we stored the label in a memory box, **again** `\setbox1=\vtop{\bigbf\hbox{Here}\hbox{again\quad}}`; notice the use of `\quad` at the end of the longest line inside the box, to ensure a reasonable amount of space between the label and the text. Then we continued exactly as above, saying `\cornerbox{\box1}{3}{6pt}` to start the paragraph. The font `\bigbf` is cmbx12 at 12 pt.

As usual, if `\cornerbox` is used inside a group, the paragraph must end before the group. Use `\par` or a blank line, if necessary.

---

[2] Pascal, *Pensée* 347, translated by W. F. Trotter.

8.9 Spacing between boxes

As we've seen several times, TEX adds some glue between boxes when it piles them up. Three variables control this behavior: `\baselineskip`, `\lineskip` and `\lineskiplimit`.

TEX first tries to arrange consecutive boxes so that their baselines are separated by `\baselineskip`; this variable is set by plain TEX to 12 pt, and in general its value should be slightly more than the size of the current font, so lines of text are harmoniously spaced.

If the box above is too deep, or the box below is too high, this rule would make the two boxes get too close to one another. In this case TEX instead separates the two boxes by the value of `\lineskip`, which is 1 pt in plain TEX. What is considered too close? The threshold is `\lineskiplimit`, which plain TEX sets to 0 pt.

Two examples will illustrate these rules:

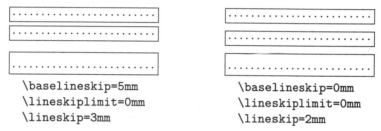

```
\baselineskip=5mm          \baselineskip=0mm
\lineskiplimit=0mm         \lineskiplimit=0mm
\lineskip=3mm              \lineskip=2mm
```

The heights of these boxes are 2, 2 and 4 mm, and their depths are all 2 mm. In the example on the left, the two first boxes are separated by $1 = 5 - (2 + 2)$ mm of spacing. This is acceptable, since it is no less than the `\lineskiplimit`. Between the second and third boxes the `\baselineskip` rule would give a skip of 5 mm − depth − height = −1 mm. This is less than the `\lineskiplimit` of 0 mm; since TEX cannot respect the `\lineskiplimit`, it places instead the `\lineskip` of 3 mm between the boxes.

On the right, on the other hand, `\baselineskip` is zero, so it's never possible to respect the `\lineskiplimit`.[3] TEX then places the `\lineskip` between all boxes.

The `\baselineskip` and the `\lineskip` can be given elasticity, but this is usually not done because even a minute difference in spacing between lines is easily picked up by the eye. For typesetting straight text, then, you should stick to rigid line spacing, and save the glue for special applications, like a table that should fill a whole page.

Turning off interline spacing

Sometimes you want to eliminate the spacing between two boxes in vertical mode; you can do this by placing `\nointerlineskip` between the two. To turn off interline spacing altogether, use `\offinterlineskip`.

---

[3] Never, as usual, turns out to be an exaggeration. Remember that boxes can have zero or negative height and depth.

In this figure, the settings of the variables are the same as in the first part of the previous one:

```
\vbox{
\box1
\nointerlineskip
\box2
\box3}
```

```
\vbox{
\box1
\box2
\nointerlineskip
\box3}
```

```
\vbox{
\offinterlineskip
\box1
\box2
\box3}
```

8.10 Rules

Rules are boxes filled with black. Here's how you get them:

```
\hrule height 2pt depth 1pt width 20in
\vrule height 20pt depth 5pt width 1pt
```

There is no backslash before `height`, `depth` and `width`. Any of these attributes may be absent, but if present they must come in this order.

There are two types of rules: `\hrule` and `\vrule`. The 'h' and 'v' stand for horizontal and vertical, as usual, but as we'll see an `\hrule` can draw a vertical line, and vice versa. The real distinction is that you can only use `\hrule` in vertical mode, and `\vrule` only in horizontal mode.

Horizontal rules

You can use `\hrule` between paragraphs, or inside a vertical box. But if you try to use it inside a paragraph, TₑX will end the paragraph and enter vertical mode. If you're in restricted horizontal mode, say inside an `\hbox`, you will simply get an error.

Here's a graphic illustration of this: we're going to bluntly say `\hrule height 1pt depth 1pt width 1in`
in the middle of this sentence. Before the rule, TₑX was in horizontal mode, setting a paragraph. After finding the rule, it changed to vertical mode, and had to finish the paragraph. Then it created the rule, which went under the previous sentence (since TₑX is in vertical mode). Finally it started another paragraph when it saw the letter 'i' after the rule.

The rules before and after this paragraph were obtained with `\medskip\hrule` `\medskip`. We had to add skips by hand because, as we've discussed, no spaces are placed before or after rules, unlike the situation with boxes.

This example also illustrates what happens when you leave out the attributes of an `\hrule`. If you leave out the `width`, the rule grows to be as wide as the immediately enclosing box. If there is no enclosing box, it grows as wide as the page (\hsize). So much for a missing `width`. The other attributes are simpler: if `height` or `depth` are missing from an \hrule , TEX gives them the default values .4 pt and 0 pt.

The box shown on the left has one rule, the first, whose width was not specified:

```
\vbox to 1.5in{\vfil
\hrule \vfil
\hrule width 16mm \vfil
\hrule width 15mm \vfil ...}
```

This rule comes out spanning the whole box. The width of the box, in turn, is determined by the lengths of the other rules, because there is no horizontal mode material (characters, etc.) in it. If there were, the width of the box would be \hsize .

Vertical rules

You can use \vrule any time you're in horizontal mode:

\hbox{\vrule height 15pt depth 5pt width 3pt} . . . . ▮

\hbox{\vrule height .4pt depth 0pt width 2cm} . . . . . . ▁▁▁▁▁▁▁▁▁▁

If you try to use \vrule in vertical mode, TEX will switch to horizontal mode and, as you know by now, give you a whole paragraph of width \hsize .

If `width` is missing for a \vrule , TEX uses the value .4 pt. If `height` or `depth` are missing, TEX uses the height and depth of the immediately enclosing horizontal box. For example,

```
\hbox{\vrule height 10pt depth 0pt\quad
      \vrule height 0pt depth 5pt\quad
      \vrule width 3pt}
```

makes the last rule as tall as the first and as deep as the second: ▕ ▕ ▐▪

Two exercises

1. Look carefully at the following code, and explain the results:

```
\vbox{\hsize=1in\parindent=0pt
  1 {\vrule height 15pt depth 2pt width 3pt} xyz
  2 \vrule\ xyz\hfil\break
  3 \vrule\ xyz
  4 \hrule\ uvw
  5 \vrule\ uvw}
```

> 1 ▌xyz 2 ▏xyz
> 3 ▏xyz 4 ▁▁▁▁▁
> ▁▁▁▁▁
> uvw 5 ▏uvw

When TEX reads the first \vrule , it is in horizontal mode, since it's just read the '1'. Thus the rule is added to the line of text, right after the '1'. Rule 2 has no dimensions specified, so TEX uses the height and depth of the line it's in. It turns out that the tallest and deepest thing on the line is rule 1, so rule 2 inherits its height and depth (the braces around the rule are irrelevant!). The width of rule 2 is not inherited from rule 1; it takes on the default value .4 pt.

The first line is ended by the `\hfil\break`, but the paragraph continues. The next line has the height of the '3' and depth of the 'y', and those are the dimensions imparted to rule 3. This line is separated from the previous one by the normal interline glue.

Rule 4 is an `\hrule`, and puts TEX in vertical mode. The width not having been set, it defaults to the width of the enclosing vertical box, which is the current value of `\hsize` (since there are paragraphs in the box). The height and depth are the default .4 pt and 0 pt.

After rule 4, TEX goes into horizontal mode again as soon as it sees the 'u'. This line has again the height of a digit; but it has depth zero, since there are no descenders. This explains the height and depth of rule 5. Rules 3 and 5 touch rule 4 because no vertical spacing is added above or below a horizontal rule.

2. Produce the two patterns below. The rules range in length from 2 mm to 20 mm, in increments of 2 mm; and they're separated by 2 mm of vertical spacing.

Let's see how the pattern on the left is obtained. It's made up of three boxes strung together horizontally, and pushed together as necessary; on the right, the same three boxes are not pushed together so much.

```
\hbox to 40mm{$\vcenter{\box1}\hss\vcenter{\box2}
  \hss\vcenter{\box3}$}
```

By now you know how to make the first box:

```
\setbox1=\vtop{\hrule width 20mm\kern 2mm
  \hrule width 18mm\kern 2mm...}
```

Box number 2 contains the TEX logo in 30 point Times Roman:

```
\setbox2=\hbox{\font\times=Times at 30pt\times\TeX}
```

Box number 3 is a little bit harder—how to make a rule of length 18 mm sit on the right side of the box, rather than on the left? We must put horizontal glue to the left of the rule somehow, which means we must go into horizontal mode. Setting the `\hsize` to 20 mm and saying `\hfill\hrule width 18mm\par` almost works, but not quite: `\hrule` is not allowed in horizontal mode. We must wrap the `\hrule` in a `\vbox`, so it is read in vertical mode. Also, we must say

`\offinterlineskip` because now we're stacking boxes, rather than rules, and we don't want the interline glue to disturb the spacing.

```
\setbox1=\vtop{\offinterlineskip\hsize 2cm
    \hfill\vbox{\hrule width 20mm}\kern 2mm
    \hfill\vbox{\hrule width 18mm}\kern 2mm...}
```

There are many other possibilities. Here are two:

• Replacing `\vbox{\hrule width 18mm}` by `\vrule height .4pt depth 0pt width 18mm`, we effectively create a horizontal rule with |.

• Replacing `\hfill` by `\moveright 2mm` (for the 18 mm rule) we avoid the need to go into horizontal mode and to set `\hsize`.

8.11 More practical examples

Framing a box

Many figures in this chapter show the contents of a box surrounded by a frame. They're created with, e.g., `\boxit{2pt}{...}`; this surrounds the stuff indicated by `...` with 2 pt of white space on all sides, followed by a frame of thickness equal to .4 pt.

```
\def\boxit#1#2{\hbox{\vrule
    \vtop{%
        \vbox{\hrule\kern#1%
            \hbox{\kern#1#2\kern#1}}%
        \kern#1\hrule}%
    \vrule}}
```

To figure out how `\boxit` works, we look at its commands from the inside out, in the order in which they are executed. First, `\hbox{\kern#1#2\kern#1}` puts the desired amount of spacing, represented by `#1`, to the left and to the right of the material represented by `#2`. In the next step, `\vbox{\hrule\kern#1...}` adds spacing above the resulting box, and also the top of the frame. The same is done at the top by `\vtop{...\kern#1\hrule}`; finally, `\hbox{\vrule...\vrule}` draws the sides of the frame.

(Notice the `%` at the end of the lines; they are there to avoid the carriage returns being interpreted as blanks, while we are in horizontal mode. Take them away and check what happens.)

Compare this with the code for `\emptybox` in section 8.8. Why is the baseline preserved through the steps above? What happens if we add the horizontal kerns together with the `\vrule`s in the outermost step, rather than in the innermost?

 To get a double frame, it is enough to use `\boxit` twice: `\boxit{2pt}{\boxit{2pt}{...}}`.

Drawing box outlines

To draw just the outline of a box, you can use the `\emptybox` macro of section 8.8 as the second argument to `\boxit`. But throughout this chapter we've used a somewhat more complicated macro, called `\drawbox`, which shows not only the outline but also the baseline. Its definition illustrates most of the fundamental concepts we've been discussing.

```
\def\drawbox#1#2#3{%
  \setbox1=\vbox{\hrule\hbox to#3{\vrule height#1\hfil\vrule}}%
  \setbox2=\hbox to#3{\vrule \dotfill\vrule}%
  \setbox3=\vtop{\hbox to#3{\vrule depth#2\hfil\vrule}\hrule}%
  \vbox{\offinterlineskip\box1\box2\box3}}
```

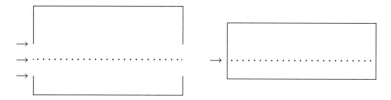

As you can see in the figure, our prototype box is obtained by stacking three pieces. The upper piece, a `\vbox`, is built in two steps. We first draw two vertical rules separated by the width of the desired box:

$$\texttt{\\hbox to \#3mm\{\\vrule height\#1mm\\hfil\\vrule\}}$$

We then add a horizontal rule at the top: `\setbox2=\vbox{\hrule...}`. Since the vertical rules have no depth, the baseline of the `\vbox` is all the way at the bottom.

The bottom piece is built along the same lines; to put the baseline at the top, we set the height of the vertical rules to zero and put them in a `\vtop`. The middle piece is just the dots. The three pieces must be put together without any spacing; hence the `\offinterlineskip`.

To conclude, here's another definition:

$$\texttt{\\def\\refpoint\{\\llap\{\\lower2.5pt\\hbox\{\$\\longrightarrow\$\}\\ \}\}}$$

To show the reference point of a box, we just say

$$\texttt{\\hbox\{\\refpoint\\box10\}}$$

8.12 For the aspiring wizard

Splitting a box

How can you typeset in two columns? We won't give the solution here, but the command `\vsplit` is a step in this direction. Suppose you have a vertical box, say box 2, which is tall and narrow. The command

$$\texttt{\\setbox1=\\vsplit2 to 1in}$$

splits off the first inch of `\box2` and puts it into `\box1`. For the example below, we filled box 2 with

```
\setbox2=\vbox{\hsize 48mm\parindent=0pt
   \raggedright\eighttrm\baselineskip=10pt
   The ...
}
```

and split its top half into box 1 with

```
\setbox1=\vsplit2 to .5\ht2
```

Here's the result; we've put frames 3 pt outside the box boundaries, so you can visualize the boxes more easily. The actual command we wrote was `\centerline {\boxit{3pt}{\box1}\hfil\boxit{3pt}{\box2}}`:

> The derailleur is operated by levers and cables and springs somewhat like those of caliper brakes. It changes gears by lifting the chain from cog to cog (rear) or chainring to chainring (front). It is an ancient system, obscurely conceived in the

> 19th century, championed by Velocio, endlessly refined, and full of compromises. The chain-line is imperfect, and owing to this, certain gears are best avoided. And components that should apparently work together frequently don't.[4]

Something isn't quite right: there is extra space at the bottom of box 1 and at the top of box 2, and TₑX complains about an underfull `\vbox`. The first problem is that the height of box 1 was prescribed exactly; TₑX filled it with as many lines as would fit, but there was a bit of space left over, and no stretchability to make it up. This leftover space at the end of box 1 can be eliminated if we say `\setbox1=\vbox{\unvbox\vsplit2 to .5\ht2}`; to avoid the underfull box report the only solution is to temporarily set `\vbadness=10000`.

The space at the top of box 2 is trickier. It comes from a variable `\splittopskip`, set by plain TₑX to 12 pt. In fact, this variable is there to help the alignment—the idea is that the first baseline in the beheaded box should be at a known distance from the top, regardless of how high the first line happens to be. TₑX arranges for this distance to be `\splittopskip` by inserting glue to make up for the difference between that and the height of the first line. If the height is greater than `\splittopskip`, no glue is inserted. In our case, the height of the line was 5.56 pt, so the glue inserted in consequence of the `\splittopskip` was 4.44 pt.

The problem is that at the top of the original box no such glue is added. One would be tempted to set the `\splittopskip` to zero, so both first lines would be right at the top of the respective boxes; and indeed this would work nicely here, where both lines happen to have the same height. But if they didn't, their baselines would be at different distances from the top, and it would be hard to align the two chunks of text side by side. A better solution is to use a strut, or invisible box, when setting the original box, so the height of its first line is exactly equal to the `\splittopskip`.

---

[4] Bill Walton and Bjarne Rostaing, *Total Book of Bicycling*, Bantam, 1985.

It is also a good idea to make the \splittopskip equal to the \baselineskip, as plain T_EX does:

```
\setbox2=\vbox{\hsize 48mm\parindent=0pt
   \raggedright\eightrm\baselineskip=10pt
   \splittopskip=\baselineskip
   \leavevmode\vbox to\splittopskip{}%
   The ... }
\setbox1=\vbox{\unvbox\vsplit2 to .5\ht2}
```

Now the result is impeccable:

The derailleur is operated by levers and cables and springs somewhat like those of caliper brakes. It changes gears by lifting the chain from cog to cog (rear) or chainring to chainring (front). It is an ancient system, obscurely conceived in the

19th century, championed by Velocio, endlessly refined, and full of compromises. The chain-line is imperfect, and owing to this, certain gears are best avoided. And components that should apparently work together frequently don't.

The truth about \vcenter

In this chapter we glossed over an important difference between \vcenter, on the one hand, and \vbox and \vtop, on the other. We know that \vcenter makes a centered box, but what does that mean?

It turns out that when T_EX adds such a box to the current math formula (recall that \vcenter can only be used in math mode), it shifts it so that the box's centerline coincides with the *axis*. The axis is an imaginary horizontal line that is defined in math mode only: it is where the minus sign $-$, for example, is placed. Many other symbols, like parentheses, are placed symmetrically about the axis: $(-)$.

The height of the axis—the distance to the baseline—depends on the current math symbol font. For plain T_EX's ten-point fonts, it equals 2.5 pt. It follows that in the construction \hbox{$\vcenter{...}$}, which we used several times this chapter, the baseline of the resulting box is *not* in the middle, but 2.5 pt below it. In the arrangement

(the third diagram in section 8.8) this was exactly what we wanted: the picture would look wrong if the the baseline bisected the box in the middle, because it runs along the bottom of the line of text.

When you align several \vcenters side by side you don't have to worry about this shift, because relative to one another they're all at the same level. This, too, was exploited in section 8.8 (second diagram).

Nonetheless, there may be cases when you do want to place a \vcenter in such a way that its centerline matches the baseline of adjacent boxes. In such a case you can use the \lower command:

The center of this ◇ is on the baseline.

```
\centerline{The center of this
  \lower2.5pt\hbox{$\vcenter{\hbox{$\diamondsuit$}}$}
is on the baseline.}
```

(This is essentially what we did in the definition of \refpoint, at the end of section 8.11; a \longrightarrow doesn't have to be placed in a \vcenter, because it's already centered about the axis.)

One more thing: instead of explicitly typing 2.5pt, you can say

```
\fontdimen22\textfont2
```

This will give the right axis height no matter what, whereas an explicit dimension won't work if you switch to a larger font, for example.

9
Alignments

Alignments, or tables, are one of the least pleasant parts of TeX. This is because there is an enormous amount of variation among tables, and TeX must be flexible enough to handle the whole spectrum. This flexibility comes at a price: you must tell TeX quite explicitly what you want your table to look like.[1] This doesn't mean that making tables is hard, just that there are lots of little things that can go wrong, and you may have to give it a couple of shots before your tables come out the way you want. So don't get discouraged—if you are familiar with the basic concepts that we've discussed so far, this chapter may well promote you to the rank of TeX Masters.

The first step in typesetting a table is to "explode" it in your mind into components. Most often the table is made of rows, each with a certain number of entries: corresponding entries in each row align vertically in columns, and are closely related, both logically and typographically. Here is a fairly run-of-the-mill table, together with its "exploded view:"

name	type	value
x	*integer*	1987
y	*real*	3.14159
z	*boolean*	false

name	type	value
x	*integer*	1987
y	*real*	3.14159
z	*boolean*	false

---

[1] In plain TeX, that is. LaTeX, or any of several existing table-making macro packages, will give you "higher-level" commands that are adequate for most purposes. See also plain TeX's tabbing facility, described in chapter 10.

9.1 The preamble, a.k.a. recipe

After this mental decomposition, you're ready to give TeX the information it needs: *How* should the rows be typeset, and *What with*? You do this with the `\halign` command, based on this skeleton:

```
\halign{
    ..#.. & ..#.. & ..#.. & ..#.. & ..#.. \cr
    ..... & ..... & ..... & ..... & ..... \cr
    ........................................
    ..... & ..... & ..... & ..... & ..... \cr
}
```

The material in braces after the `\halign` control sequence is divided into rows, each terminated by `\cr`; entries within each row are separated by ampersands `&`. For the most part, these rows correspond to the rows of the table, but the first "row" of the `\halign` is special: it is called the preamble, and it's there that you answer the first question above: How should each row be set?

Notice that the preamble contains sharp signs `#`, alternating with ampersands. Everything around a `#` and between the flanking ampersands is material common to all the entries in a given column. The `#` itself represents something that will be plugged in from the other rows.

The rest of the `\halign` answers the second question, What should the rows be filled with? As it reads each row, TeX chops it into individual entries at the ampersands. It then splices each entry into the corresponding entry of the preamble, or template, in place of the `#`. The result of the splicing is what TeX typesets as the table entry.

9.2 Simple alignments

By a simple alignment we mean one with no exceptional cases and no rules separating rows and columns. Our first alignment will have simplest possible preamble:

```
\halign{
    #   &   #   &   #    \cr
    name&type    &value  \cr
    $x$ &integer&1987    \cr
    $y$ &real    &3.14159\cr
    $z$ &boolean&false   \cr
}
```

name	type	value
x	integer	1987
y	real	3.14159
z	boolean	false

The three templates here consist of nothing but a `#`. To build the first row of the output, TeX replaces the first `#` in the preamble by `name`, the second by `type`, and the third by `value`. It proceeds similarly for the other three rows.

This does the job, but it's ugly! There isn't enough space between the columns. TeX makes the a column exactly as wide as its widest entry, and goes on to place the columns side by side, without any glue between them. So in fact, there would have been no space at all separating name and type but for the fact that the templates contain spaces. It turns out that spaces at the beginning of a template or row entry are discarded, so what TEX is really typesetting is `name` SP, `type` SP, and so on. (Remember that several consecutive spaces are collapsed to one.)

To improve the situation, we put some `\quad` s in the preamble, and italicize the second column. Also, from now on, we place our ampersands right after the previous entry, so no spurious blanks will creep in. (Of course, we don't have to align the columns at all, but doing so helps make the source file more intelligible by highlighting the structure of the table.)

```
\halign{
  #&        \quad\it#& \quad #\cr
  name&     type&      value\cr
  $x$&      integer&   1987\cr
  $y$&      real&      3.14159\cr
  $z$&      boolean&   false\cr
}
```

name	*type*	value
x	*integer*	1987
y	*real*	3.14159
z	*boolean*	false

Now each entry in the second column is effectively preceded by `\quad\it`, and each entry in the third column by `\quad` ! That's why templates are so useful. They let you "factor out" the commonalities of each column, making it possible to change the appearance of the whole table without modifying its entries.

Notice that the `\it #` in the second template is not surrounded by braces. It doesn't have to be: alignment entries constitute groups in themselves, so font changes and assignments made inside them are local by default. By the same token, groups can't straddle entries:

    Wrong: $... & {\bf... & ...}... & ...$
    Right: $...$ & ${\bf...}$ & ${\bf...}...$ & $...$

(But what counts is the whole alignment entry, after the value of # has been plugged in, so a left brace in the preamble can be matched by a right brace in the body of the table.)

We've seen that TEX makes each column as wide as its widest entry. What then of the other entries? Basically, any glue that might be present in them is stretched so the width of the entry matches the column width. In the latest version of our table, the entries had no stretchability at all, so they appeared flush left in underfull boxes. This is harmless; TEX doesn't complain of underfull boxes inside alignments.

This process makes it very easy to specify right-aligned or centered columns: it's enough to place appropriate springs in the preamble.

```
\halign{
  \hfil#\hfil&\quad\it#\hfil&
                \hfil\quad#\cr
  \bf name&  \bf type&  \bf value\cr
  $x$&       integer&   1987\cr
  $y$&       real&      3.14159\cr
  $z$&       boolean&   false\cr
}
```

name	**type**	**value**
x	*integer*	1987
y	*real*	3.14159
z	*boolean*	false

(If a row of your \halign is too long to fit in a single line, it's best to break it right after an ampersand, for the same reason that you can leave spaces after, but not before, an ampersand.)

The combination \hfil\quad occurs so frequently that we put an abbreviation for it in the macros file read in at the beginning of every run (cf. section 1.7): \def\hfq{\hfil\quad}. You should do the same with any construction that you find yourself using very often: it saves time and decreases the probability of error.

9.3 Some practical suggestions

The \halign command can only be used in vertical mode, since it creates a stack of horizontal boxes (the rows). As usual when stacking up boxes, TeX adds interline glue between them, which is why in the table of section 9.2 the rows came out nicely spaced as if they were lines in a paragraph. Also as usual, TeX feels free to break the stack between pages if it's working in ordinary vertical mode (that is, not inside any boxes). To avoid this, or to set a table in horizontal mode, you can wrap it in a vertical box: \vbox{\halign{...}}.

There is one exception to the vertical mode rule: An \halign can be used all by itself in display math mode, that is, between double dollar signs $$. However, this places the alignment flush left on the page, which is almost never what you want to do, so this construction is rare. Much more common is to say

$$\vbox{\halign{...}}$$

which centers the table horizontally. This is perhaps the most convenient way to center tables, but there are many others, including one that avoids the need for a \vbox and the consequent impossibility of breaking the table across pages. For details, see \tabskip in section 9.12.

Whatever you do with the \halign, it's best to start coding it from the outside, typing in a skeleton first:

$$\vbox{\halign{

}}$$

Only then should you fill in the rows. In our experience, when you don't do this you have a better than even chance of forgetting one or both of the closing braces. The result is that everything from there on is seen by TeX as part of the alignment, and you get an error like

! You can't use '\end' in internal vertical mode.

Another common error consists in forgetting the \cr at the end of the last line. Certain macros, like \matrix, let you get away with it, because they use a magic control sequence \crcr that compensates for the omission (section 12.10); but \halign is unforgiving.

9.4 Treating special cases

A shortcut for the preamble

Values of x:	0	1	2	3	4	5	6	7	8	9
Values of x^2:	0	1	4	9	16	25	36	49	64	81

This table has eleven columns; apart from the first, they all conform to the model `\hfil#\quad`. To avoid repeating this template ten times, we can use a shortcut:

```
\halign{
#\hfil\quad&&\hfil#\quad\cr
Values of $x$:    & 0& 1& 2& 3&  4&  5&  6&  7&  8&  9\cr
Values of $x^2$:  & 0& 1& 4& 9& 16& 25& 36& 49& 64& 81\cr
}
```

Generally, # and & must alternate in the preamble, with a # preceding the first &. The shortcut is to put in an extra & just before one of the templates; this causes the portion of the preamble following of the irregularity to be repeated as many times as necessary to account for all the columns in the table. A preamble of the form

$$A \text{ \&\& } B \text{ \textbackslash cr},$$

where # and & alternate inside A and B, is equivalent to

$$A \text{ \& } B \text{ \& } B \text{ \& } B \ldots \text{ \textbackslash cr}$$

for as long as necessary, and similarly for a preamble of the from `&`B`\cr`.

Here is a common application—it is used, for example, in the `\matrix` macro (section 11.25). We want to make a table all of whose columns are separated by `\quad` and, say, left-justified. If we make the preamble `&#\hfil\quad\cr`, we get an extra `\quad` after the last column, and if we make it `&\quad#\hfil\cr`, it's the first column that gets a spurious `\quad`. Either way the extra spacing shows when the table is surrounded by a frame or by parentheses (as matrices often are). The right solution is `#\hfil&&\quad#\hfil\cr`: the first column has no spacing, and each subsequent one is separated from the preceding one by a `\quad`.

Empty entries

TEX won't raise an eyebrow if a row has fewer entries than the preamble—it just skips the missing entries just before `\cr`. But watch out: `toto&&&&&&\cr` is not the same as `toto\cr`. In the first case the corresponding templates are still used, with # replaced by nothing; in the second, the templates are skipped.

```
\halign{
$#$&&\hfil\quad$#$\cr
1\cr
1&1\cr
1&2&1\cr
1&3&3&1\cr
1&4&6&4&1\cr
1&5&10&10&5&1\cr
}
```

1					
1	1				
1	2	1			
1	3	3	1		
1	4	6	4	1	
1	5	10	10	5	1

Skipping templates

In the example table of section 9.2, all but one entry in the first column are in math mode. It would be nice to further simplify the code by writing `\hfil$#$\hfil` for the corresponding template, taking the dollar signs out of the individual entries.

As usual, there is a way to deal with the recalcitrant exception: if a table entry says `\omit` at the very beginning, TEX ignores the corresponding template, this time only:

```
\halign{
\hfil$#$\hfil&\quad\it#\hfil&
                  \hfil\quad#\cr
\omit\bf name&\bf type& \bf value\cr
x&              integer&  1987\cr
y&              real&     3.14159\cr
z&              boolean&  false\cr
}
```

name	type	value
x	*integer*	1987
y	*real*	3.14159
z	*boolean*	false

You may be wondering if the change was worth the effort—we didn't save any keystrokes. But by isolating a common feature of all or most entries in a column, we make it easier to change that feature later, if necessary. We also make the table structure clearer.

Exercise

$$T_EX \qquad T_EX \qquad T_EX \qquad T_EX$$
$$T_EX \quad T_EX \qquad T_EX \quad T_EX \qquad T_EX \quad T_EX$$
$$T_EX \qquad T_EX \qquad T_EX \qquad T_EX \qquad T_EX$$

Typeset this frieze as a three-column alignment with body entries as short as possible. For example, the middle row should read `8&8&8\cr` if the TEX's at the middle of each V are separated by 8 pt. (They're separated by 35 pt at the mouth, and the Vs themselves are separated by two quads.) Don't read any further until you've tried it!

Here's one way to do the preamble, with repeated templates that split the inter-column spacing equally between left and right (remember that `\hfq` stands for `\hfil\quad`):

$$\&\hfq\TeX\hskip\#pt\TeX\hfq\cr$$

These templates are fine except for the tips of the Vs. The first row will read `...&35&...\cr` and the third `35&...&35\cr`. For the tip, we use `\omit` to avoid the template and put in a single centered TEX:

```
\halign{
&\hfq\TeX\hskip#pt\TeX\hfq\cr
\omit\hfq\TeX\hfq&        35&      \omit\hfq\TeX\hfq\cr
      8&                  8&            8\cr
      35&      \omit\hfq\TeX\hfq&       35\cr
}
```

It is very important that `\omit` be right at the beginning of the entry (after blanks, which are ignored): TEX is on the lookout for it then and only then. If anything else is found, it will be plugged into the template.

9.5 Excessively wide entries

Work	Year of Publication
Montesquieu's *Considérations*	1734
Voltaire's *Essai sur les Mœurs*	1745
Hume's *History of England*	1754
Gibbon's *Decline and Fall*	1776

In the table above, with preamble `#\hfil\qquad&\hfil#\cr`, the top entry of the right-hand column is much wider than the others, so TEX's default behavior—letting this entry control the width of the whole column—makes the spacing excessive. To balance the table better, we'd like to make the long entry "spill over" into the neighboring column.

The idea is to fool TEX into thinking that the entry is not the widest, by giving it a negative width. Any negative width will do, since any negative number is less than a positive one. To be on the safe side, we add −1000 pt of spacing to the left of the entry; remember that there is an `\hfil` there already, as part of the template, so the excess negative glue will be canceled out:

```
\bf Work& \hskip -1000pt \bf Year of Publication\cr
```

Work	Year of Publication
Montesquieu's *Considérations*	1734
Voltaire's *Essai sur les Mœurs*	1745
Hume's *History of England*	1754
Gibbon's *Decline and Fall*	1776

Plain TEX's `\hidewidth` macro officializes this idea: it combines the `\hskip` `-1000pt` with an `\hfill`, so you can use it whether or not the preamble contains a spring. Consequently, if you put `\hidewidth` at the beginning of an entry, that entry is allowed to spill over into the column to its left, and if you put it at the end, the entry spills over to the right.

9.6 Inserting material between rows

The construction `...\cr\noalign{...}`, where the material in braces is anything that is allowed in vertical mode, inserts that material between the rows of an alignment. You can have several `\noalign`s one after the other, but they all must come immediately after the `\cr` that terminates the previous row (or the preamble): they doesn't make sense anywhere else.

As an example, let's improve our favorite table some more by using horizontal rules to separate the row of titles from the rest of the table.

```
\halign{
  \hfil$#$\hfil&\quad\it#\hfil&
                  \hfil\quad#\cr
  \noalign{\hrule\smallskip}
  \omit\bf name&\bf type& \bf value\cr
  \noalign{\smallskip\hrule\smallskip}
  x&            integer&  1987\cr
  y&            real&     3.14159\cr
  z&            boolean&  false\cr
  \noalign{\smallskip\hrule}
}
```

name	type	value
x	*integer*	1987
y	*real*	3.14159
z	*boolean*	false

The \smallskip s are necessary because TeX doesn't add interline glue above and below a rule. The \hrule s are exactly long enough to span the alignment, because we didn't specify their length. It's as if the alignment were a containing \vbox . But other types of material, that have an intrinsic width, will not conform to the alignment; they simply stack up with the rows, aligned on the left.

To understand that, consider the following attempt to create a centered alignment:

```
$$\vbox{\halign{
  #\hfil&&\quad#\hfil\cr
  auk&          bobolink& cassowary& dodo&     egret\cr
  asparagus& broccoli& celery& daikon*& eggplant\cr
  \noalign{\smallskip\hrule\smallskip}
  \noalign{*Japanese radish.}
}}$$
```

It fails rather miserably:

auk	bobolink	cassowary	dodo	egret
asparagus	broccoli	celery	daikon*	eggplant

*Japanese radish.

What happened? Let's retrace TeX's steps. In vertical mode, TeX encountered the alignment and set two lines of it; it then came to the \noalign s, which it set independently of the alignment. The vertical glue and horizontal rule are vertical mode material that doesn't affect the width; but when it came to the note, TeX had to start a paragraph! As you know, the resulting lines are of width \hsize , so that TeX is effectively stacking up, aligned on the left, a relatively narrow table and a line of full width. No wonder the result appears not to be centered—there is no space left over to center it!

The right way to fix this situation is always to wrap your text in a horizontal box: \noalign{\hbox{*Japanese radish.}} The width of the enclosing vertical box will then be the width of the alignment or the width of the \hbox inserted with \noalign , whichever is greater.

We will make repeated use of \noalign in the following sections. The Dictionary also contains other applications.

9.7 Combining columns

The construction `\multispan` n `{...}` makes the material in braces span the n next columns of an `\halign`. The templates for those columns are ignored, as if all the individual entries had started with `\omit`. (In fact they do: `\multispan` is a macro that puts `\omit`s in the right places.) In the table

STRASBOURG MARKET

Item	Origin	Price per kg	Weight
Artichokes	St. Pol de Léon	9.40F	100kg
Apricots	..............*not here yet*..............		
Kiwis	New Zealand	14.00F	30kg

We combined all four entries of the first row into one (the title), and also the last three entries of the fourth row. Here's the code we used:

```
\halign{
  #\hfil&     \quad #\hfil&       \hfil\quad#& \hfil\quad#\cr
  \multispan4\hfil\bf STRASBOURG MARKET\hfil\cr
  \noalign{\medskip}
  \bf Item&  \bf Origin& {\bf Price per kg}& \bf Weight\cr
  \noalign{\smallskip}
  Artichokes&St.~Pol de L\'eon& 9.40F&        100kg\cr
  Apricots&\multispan3\quad\it\dotfill not here yet\dotfill\cr
  Kiwis&     New Zealand&       14.00F&       30kg\cr
}
```

The `\multispan4{...}` command appears at the beginning of the first row of the `\halign`'s body, so it replaces entries 1 through 4. All the corresponding templates were discarded, but they were no good anyway, since we wanted the title centered.

The `\multispan3{...}` is analogous. It appears right after the first `&` of the fourth row, that is, in lieu of entries 2 through 4. Again the templates were skipped, but the `\quad` at the beginning at the second template is essential to keep the alignment, so we had to copy it over into the entry. Notice also the use of `\dotfill`, instead of `\hfil`, on both sides of the text.

Here are some things to keep in mind:

- When you write `\multispan3`, you're merging three entries, so you'll be skipping only two ampersands.

- `\multispan`, just like `\omit`, must come at the beginning of an entry—otherwise TEX gets the order to skip the template after having started to use it, and goes into a tail spin.

- If you leave a space after `\multispan3`, it will go through to the output, and will be noticeable if your entry is left-justified or starts with `\dotfill` and the like. Write `\multispan3\dotfill` instead.

- If you're spanning more than nine columns, the number should go in brackets: `\multispan{14}`. Here too, a space after the braces will appear on the output.

- `\multispan1` is synonymous with `\omit`.

9.8 Aligning digits

Item	**Price** (F/kg)	**Weight** (kg)	**Total** (F)
Artichokes	9.40	100	940.00
Apricots	7.30	12	87.60
Kiwis	14.00	30	420.00
Grand Total	. . . .		1467.60 F

Notice carefully the alignment of digits in this table. As a group, each column of figures is centered with respect to the column title; but individually, the figures are right-justified, so their decimal points align. You may enjoy trying to puzzle out how to achieve this.

In the event we used a trick: we fooled TeX into thinking that all figures in the same column have the same width by padding them with an "invisible digit" defined like this:

```
\catcode'\*=\active      \def*{\hphantom{0}}
```

The first of these makes * into a macro (see section 12.6), and the second gives it a meaning, making it stand in for a digit (section 11.17). These changes are confined by the `\vbox` that encloses the alignment. Here's the rest of the code:

```
\halign{#\hfil&&\hfil\quad #\hfil\cr
    \bf Item&    {\bf Price} (F/kg)&
       {\bf Weight} (kg)&    {\bf Total} (F)\cr
    \noalign{\smallskip}
    Artichokes& *9.40& 100&  *940.00\cr
    Apricots&   *7.30& *12&  **87.60\cr
    Kiwis&      14.00& *30&  *420.00\cr
    \noalign{\smallskip}
    \multispan3{\bf Grand Total}\quad\dotfill&
                        1467.60\rlap{ F}\cr
    }
```

We couldn't resist the temptation of showing off another trick: the use of `\rlap` to place a space and an 'F' to the right of the grand total, without disturbing the alignment of the decimal points. The width of `\rlap`'s argument is neutralized by a negative spring, so it doesn't count toward the width of the entry—it's as if TeX weren't aware that it had written the stuff at all!

The * trick is also useful when the entries have different numbers of decimal places and must be aligned by their decimal points, as is conventional. The little table on the right was obtained by typing

```
\halign{#\cr *44.1*\cr 172.**\cr **0.12\cr}
```

44.1
172.
0.12

9.9 Horizontal rules and spacing

The sections from here till the end of the chapter are a bit more difficult than what we've seen so far. You may want to just skim through them the first time around, just to see what's possible to do. Later you can refer back to them as needed.

In section 9.6 we used \noalign to separate rows of an alignment with horizontal rules. We ran into the need to add spacing above and below the rule, because TEX doesn't do so automatically. The solution we used there, interspersing \smallskip s, leaves something to be desired, fixing as it does the distance between the bottom of a line and the following rule, and between the rule and the top of the next line. It is generally desirable instead to have a uniform distance between baselines and rules. In our example table, the distance between baseline and rule in the first row is greater than in the last row, because the first row has letters with descenders, like 'y', while the last one doesn't.

A general approach to solve this problem is based on the important idea of a strut, which we mentioned briefly in section 8.12. A strut is something invisible, but fairly tall and deep, so it "sets the pace" for the line that it's on. If every line contained a strut, no interline glue at all would be needed to separate them, because the height and depth the lines inherit from the strut would make their baselines be separated by a fixed amount of space. (This is true only if the struts are the tallest and deepest components on their lines, as is normally the case. But if you have complicated formulas with fractions or big subscripts, this condition may no longer hold, and things get more complicated.)

Plain TEX defines the \strut macro as (basically) a rule of width zero, height 8.5 pt and depth 3.5 pt. This means that if two consecutive rows of an alignment have struts and there is no interline glue, their baselines are 12 pt apart: which is exactly the normal value of \baselineskip .

```
\vbox{\offinterlineskip\halign{
  \strut#&\hfil$#$\hfil&
   \quad\it#\hfil&\hfil\quad#\cr
  \noalign{\hrule}
  &\omit\bf name&\bf type&\bf value\cr
  \noalign{\hrule}
  &x&            integer& 1987\cr
  &y&            real&    3.14159\cr
  &z&            boolean& false\cr
  \noalign{\hrule}
}}
```

name	type	value
x	*integer*	1987
y	*real*	3.14159
z	*boolean*	false

Here \offinterlineskip was used to turn off the interline glue within the \vbox , and a \strut was placed in the preamble, so as to be replicated in every row of the alignment. Now the distance from the baseline of the first row to the following rule is now exactly the same as for the last row. The strut would have worked in any column, but we put it in a column by itself so it's independent of the other entries: if it were in the name column, for instance, it would have to be copied over into the entry that starts with \omit .

For most tastes, the amount of spacing between baseline and rules in this table is insufficient. To increase it, we could redefine `\strut` to be taller and deeper:

```
\vbox{\offinterlineskip
  \def\strut{\vrule height 10.5pt
    depth 5.5pt width 0pt}
  \halign{
    ... no changes here
}}
```

name	type	value
x	*integer*	1987
y	*real*	3.14159
z	*boolean*	false

(Notice that the new definition of `\strut`, like the use of `\offinterlineskip`, is local to the `\vbox`, so it will go away after its job is done.) This works rather well when all the rows are separated by rules, but in this case it leads to the opposite problem: the rows not separated by rules are too far apart. The best solution seems to be a hybrid one:

```
\vbox{\offinterlineskip\halign{
  \strut#&\hfil$#$\hfil&\quad\it#\hfil&
         \hfil\quad#\cr
  \noalign{\hrule\vskip 2pt}
  \omit\bf name& \bf type& \bf value\cr
  \noalign{\vskip 2pt\hrule\vskip 2pt}
  x&              integer&  1987\cr
  y&              real&     3.14159\cr
  z&              boolean&  false\cr
  \noalign{\vskip 2pt\hrule}
}}
```

name	type	value
x	*integer*	1987
y	*real*	3.14159
z	*boolean*	false

In a different vein, plain TEX offers an `\openup` macro to increase the spacing between rows of an alignment for which interline spacing has not been turned off. This is most useful in display math mode, with the `\eqalign` macro and its relatives. By saying

$$\texttt{\{\textbackslash openup 3pt\textbackslash halign\{...\}\}}$$

you effectively increase the `\baselineskip` by 3 pt. To have effect, `\openup` should be outside the alignment: `\halign{\openup 3pt ...}` won't do any good. On the other hand, it should be confined by some group—perhaps the `\vbox` containing the alignment—or it will interfere with interline spacing in normal text.

To summarize, then, there are several ways to open up a table:

• `\noalign{\vskip...}` lets you control the spacing between individual rows, and acts just the same whether or not interline spacing has been turned off using `\offinterlineskip`.

• `\strut` in the preamble, together with `\offinterlineskip` before the table, uniformizes the height and depth of all rows, and the spacing between baselines. This combination is especially useful when there are rules between rows.

• `\openup` before the table changes the spacing between baselines for all rows. It normally makes sense only if you're not using `\offinterlineskip` (but see also section 9.11).

Rules across columns

We now know how to use `\noalign` to place rules across a whole alignment. How about rules that span some columns only? The idea is to treat the rule as part of an entry, like the leaders in the table of section 9.8. For one column, `\omit\hrulefill` works. For three, say, you can repeat `\omit\hrulefill` three times, or use the shorthand `\multispan3\hrulefill`.

Keep in mind that these short rules, being part of regular rows, are put in boxes before being stacked up. This means that to get any sort of sensible spacing you must turn off the automatic interline spacing by saying `\offinterlineskip`, and then use struts to manage the spacing yourself. Here's a typical example that you should study closely:

```
\vbox{\offinterlineskip\def\hfn{\hfil\enspace}
  \def\strut{\vrule height9pt depth3pt width0pt}
  \halign{
    \hfn#\hfn&\strut\hfn#\hfn&\hfn#\hfn\cr
    &\omit\hrulefill&    \cr
    &            T&    \cr
    \multispan3\hrulefill\cr
    T&            E&  X\cr
    \multispan3\hrulefill\cr
    &            X&    \cr
    &\omit\hrulefill&    \cr
}}
```

To make sure you understand, explain what goes wrong when each of the following changes is made:

\offinterlineskip
and \strut removed

\strut removed

\strut moved
to first column

9.10 Vertical rules

The easiest way to make vertical rules in an `\halign` is to build them up from short pieces, each manufactured within a row. The "exploded view" on the first page of the chapter gives the idea.

This at first may sound like a cumbersome solution, but in fact it turns out to be very simple, as we already have all the ingredients in place. Any table created with `\offinterlineskip`, that is, one that relies on struts to support its structure, is ready to receive vertical rules. The trick is to place \vrule at appropriate places in the preamble. Since a \vrule whose vertical dimensions are not given expands to the height and depth of the enclosing box, this gives chunks of vertical rules that connect together seamlessly.

Here is the first table of section 9.9, with vertical rules added:

```
\vbox{\offinterlineskip\halign{
  \strut#&\vrule#\quad&
  \hfil$#$\hfil&
  \quad\vrule#\quad&
  \it#\hfil&\quad\vrule#\quad&
  \hfil#&\quad\vrule#\cr
\noalign{\hrule}
&&\omit\bf name&&
       \bf type&&\bf value&\cr
\noalign{\hrule}
&&x&& integer&& 1987&      \cr
&&y&& real&&    3.14159&   \cr
&&z&& boolean&& false&     \cr
  \noalign{\hrule}
}}
```

name	type	value
x	*integer*	1987
y	*real*	3.14159
z	*boolean*	false

The entries are unchanged! The changes are all localized in the preamble: one template was introduced for each \vrule, with the surrounding spacing. The corresponding entries in the body are all empty, so we just add ampersands as needed. Making individual columns for the \vrule s is not indispensable, especially if there are no \omit s, but is good practice because it keeps things independent.

And now, for the gran finale: The Perfect Table with which we opened this chapter. We need to add a bit of spacing above and below the horizontal rules, as we did in section 9.9. But we can't write \noalign{\vskip 2pt} anymore, for that would interrupt the vertical rules. Instead, we will insert little "rows" 2 pt tall, containing only pieces of vertical rules! The height of these mini-rules can be conveniently specified by an entry that says just height 2pt: since the \vrule in the template is immediately followed by #, what TeX sees is \vrule height 2pt. Furthermore, this has to be done only once per row—the other rules borrow the height of the tallest one. All that remains to do is to turn off the strut, et voilà:

```
\vbox{\offinterlineskip
  \def\mr{\omit&height 2pt&&&&&}
  \halign{
    ...
    \noalign{\hrule} \mr
    ...
    \mr \noalign{\hrule} \mr
    ...
    \mr \noalign{\hrule}
}}
```

name	type	value
x	*integer*	1987
y	*real*	3.14159
z	*boolean*	false

This preamble came out unusually complicated because each column is treated differently, from the typographic point of view. Often you can just use an abbreviated preamble as explained in section 9.4.

Exercise

Typeset the arrangement shown here. The small squares have sides 18 pt (not counting the thickness of the walls) and the letters sit 6 pt above the floor of their squares.

9.11 Braces and tables

Horizontal braces

The `\downbracefill` and `\upbracefill` macros, which make springy braces (section 5.8), work well in alignments:

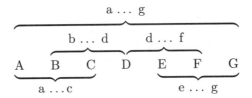

Here is the specification for this example:

```
\vbox{\offinterlineskip\openup 6pt
  \halign{&\quad#\quad\cr
    \multispan7\hfil a \dots\ g\hfil\cr
    \multispan7\quad \downbracefill \quad\cr
    &\multispan5\quad\hfil b \dots\ d\hfil
      \thinspace\hfil d \dots\ f\hfil\quad\cr
    &\multispan5\quad\downbracefill
      \thinspace\downbracefill\quad\cr
    A&B&C&D&E&F&G\cr
    \multispan3\quad\upbracefill\quad&
      &\multispan3\quad\upbracefill\quad\cr
    \multispan3\hfil a \dots c\hfil&
      &\multispan3\hfil e \dots\ g\hfil\cr
}}
```

There are several interesting points to notice:

• By saying `\offinterlineskip\openup6pt`, we've effectively made the interrow spacing always 6 pt. This is the same as making `\lineskip=6pt` and the `\lineskiplimit` so absurdly big that it can never be satisfied: see the discussion in 8.9.

• By having `\quad`s on both sides of the braces, we get them to embrace the letters for the columns they span, and no more. Without the `\quad`s they would stretch all the way to the (invisible) column boundary. The two middle braces have to be typeset as part of the same `\multispan`, since they share the 'D'.

• A similar scheme centers the labels with respect to their braces. Study the third row carefully: in order to center b ... d, we rely on the fact that `\hfil b \dots\ d\hfil` has the same width as the corresponding braces.

Plain TEX defines `\upbracefill` and `\downbracefill` in terms of `\vrule`s of unspecified height and depth—that's right, `\vrule`s—despite the fact that they stretch horizontally. For this reason, disaster ensues if you put them in the same `\hbox`, line, or alignment entry with anything that has height or depth:

```
\hbox to 1in{\upbracefill\strut} ................. ⌣███⌣███⌣
\hbox to 1in{j \downbracefill} ................... j⌣███⌣███⌣
```

If you absolutely must, you can use `\smash` to hide the height of everything else.

See section 11.19 for an example of use of vertical braces with an alignment.

9.12 Fixing the width of an alignment

Just as `\hbox to` ... lets you fix from the outside the width of a horizontal box, so too you can fix the width of a table beforehand. Again, TEX will try to satisfy your request by stretching or shrinking the available glue. But what glue? The glue inside individual entries has already been used to make them conform to the column width—it has already been "set," so to speak.

However, TEX also makes provision for glue between the columns. This glue, governed by the `\tabskip` variable, had remained on the sidelines so far, but it turns out to be a tool of great versatility.

We start with a simple alignment, with only two rows and three columns. The widest thing in each column is `\showcol`, which is defined as

```
\hbox to 1in{\leftarrowfill\hskip -1em\rightarrowfill}
```

and also serves to show where the column boundaries begins and ends.

```
\vbox{\offinterlineskip
  \halign{
    #&#&#\cr
    \noalign{\hrule\vskip 2pt}
    \strut\hfil center\hfil&left\hfil&\hfil right\cr
    \showcol&\showcol&\showcol\cr
}}
```

center	left	right
←——→	←——→	←——→

What happens when we replace `\halign` with `\halign to\hsize`? Nothing much, except that TEX complains about an underfull `\hbox`. The `\hrule` at the top stretches to length \hsize , but as we know it is not really part of the alignment. Nothing else budges, as there is no glue to stretch.

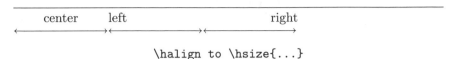

$$\text{\halign to \hsize\{...\}}$$

Next we give a non-zero value to \tabskip , but let the alignment have its natural width:

$$\text{\tabskip=.2in \halign\{...\}}$$

You can see by looking at the rule at the top, which spans the whole alignment, that TₑX puts spacing not just between the columns, but also before the first and after the last column.

So far, so good; but we still haven't been able to make the alignment the width of the page. To do that we must give the \tabskip some stretchability:

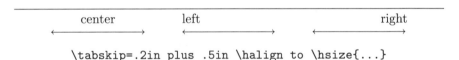

$$\text{\tabskip=.2in plus .5in \halign to \hsize\{...\}}$$

Now we're getting somewhere! How do we get rid of the spacing before the first column and after the last? It turns out that \tabskip can be changed on the fly, and the intercolumn glue responds accordingly:

- the value \tabskip happens to have when the ampersand between two templates is read is used to separate the respective columns;
- the value at the beginning of the preamble is used before the first column; and
- the value at the end of the preamble is used after the last column.

Here then is an alignment with glue between columns, but not before of after:

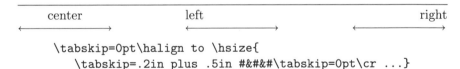

```
\tabskip=0pt\halign to \hsize{
      \tabskip=.2in plus .5in #&#&#\tabskip=0pt\cr ...}
```

Using intercolumn glue means you don't have to clutter your preambles with lots of \quad s and the like. This is especially the case if your table has vertical rules, and is another reason why vertical rules should be kept in column by themselves. Let's simplify the unwieldy preamble of the table in section 9.10:

```
\tabskip=0pt\halign{
    \strut#& \vrule#\tabskip=1em& \hfil$#$\hfil& \vrule#&
    \it#\hfil&\vrule#& \hfil## \vrule#\tabskip=0pt\cr ...}
```

(Remember that `\quad` is an abbreviation for `\hskip 1em`.) With a minimal amount of change you can now make the same table conform to a desired width:

```
\halign to .6\hsize{\tabskip=0pt
    \strut#& \vrule#\tabskip=1em plus 1in& ...}
```

gives

name	type	value
x	*integer*	1987
y	*real*	3.14159
z	*boolean*	false

9.13 Vertical alignments

Not all tables are best seen as made up of rows; some are better described as a juxtaposition of columns:

abel	basil	cecilia	desdemona
agnes	bernard	christopher	diane
amanda	bertrand	cuthbert	
anatole	brigitte		
arnold			

Admittedly, this table could be typeset with `\halign`. But imagine the hassle it would be to add acton between the first two entries of the first column! In any case, if you think of the A-words together, you should be able to code them together.

You are, of course. It probably comes as no surprise, given the 'h' in its name, that `\halign` has a vertical counterpart `\valign`. The two commands are in every way dual, so in a way you already know all there is to know about `\valign`; but since some of the consequences of this duality are far from obvious, let's look at some examples.

The code for the table above was

```
\valign{&\hbox{\strut\quad#}\cr
    abel&agnes&amanda&anatole&arnold\cr
    basil&bernard&bertrand&brigitte\cr
    cecilia&christopher&cuthbert\cr
    desdemona&diane\cr}
```

Just as the natural environment for an `\halign`, at least from TEX's point of view, is vertical mode, so a `\valign` should be used in horizontal mode. Thus, we used `\centerline` to center the alignment. Double dollar signs `$$` will do too, with the precaution of wrapping the `\valign` in an `\hbox`.

Naturally, `&` separates entries in the same column, and `\cr` separates columns. The preamble works in the same way as the preamble of an `\halign`; in this case we used the shortcut of section 9.4 to avoid repeating the same template several times.

Some essential facts about \valign

- Each entry must be enclosed in a box. Indeed, entries are typeset in vertical mode (that's right, vertical and horizontal are interchanged!), and characters floating around in vertical mode create boxes of width \hsize. Here we wrapped the template's # in an \hbox, so this takes care of all entries.

- Entries in the same column come out aligned on the left. The counterpart of this fact for \halign is that entries on the same row are aligned by their baselines, something so natural that we've taken it for granted. But here the consequence is that you can only right-align or center columns if you know their width beforehand. In that case you can say, for instance

```
\valign{&\hbox to .25\hsize{\strut\hfil#}\cr...}
```

But \hbox{\strut\hfil#}, as you know, has no effect. And placing the spring outside the box would be disastrous: it would be read in vertical mode, and there you have a box of width \hsize.

- Struts must be used to regularize the distance between rows, because no default glue from \baselineskip is added between the boxes in a column. If you find the spacing obtained with a regular strut insufficient, you can use an extra-tall strut, or use \tabskip glue, which is now added between rows:

```
{\tabskip=2pt
\valign{&\hbox{\strut\quad#}\cr
  ...
}}
```

Getting fancier

Vertical rules between columns are easy to obtain, like horizontal rules between rows in an \halign. Just insert \noalign{\vrule} after each \cr, and make sure there's enough space on both sides of the # in the template:

```
\valign{&\hbox{\strut\quad#\quad}\cr\noalign{\vrule}
   abel&agnes&amanda&anatole&arnold\cr\noalign{\vrule}
   ... \noalign{\vrule}}
```

abel	basil	cecilia	desdemona
agnes	bernard	christopher	diane
amanda	bertrand	cuthbert	
anatole	brigitte		
arnold			

You may prefer to set rules between individual entries, so they don't extend all the way down between short columns. The solution for the analogous problem in an \halign involved \omit\hrulefill. There is no \vrulefill in plain TEX, but we can easily remedy that by cribbing the definition of \hrulefill from page 357 of *The TEXbook*:

```
\def\vrulefill{\leaders\vrule\vfill}
```

The short rules connect together without the need for `\offinterlineskip`, since there is no interline glue anyway.

```
\valign{&\hbox{\strut\quad#\quad}\cr
  \multispan5\vrulefill\cr
  abel&agnes&amanda&anatole&arnold\cr
  \multispan5\vrulefill\cr
  ...
  desdemona&diane\cr
  \multispan2\vrulefill\cr}
```

abel	basil	cecilia	desdemona	
agnes	bernard	christopher	diane
amanda	bertrand	cuthbert	
anatole	brigitte		
arnold			

Horizontal rules, too, can be added by analogy. As in the preamble of an `\halign`, we alternate rows of entries with "rows" of rules, with a rule to open the procession; and we increase the number of ampersands and the arguments to `\vrulefill`, to account for the new rows.

```
\def\strut{\vrule height 10.5pt depth 5.5pt width 0pt}
\valign{\hrule#&&\hbox{\strut\quad#\quad}&\hrule#\cr
  \multispan{11}\vrulefill\cr
  &abel&&agnes&&amanda&&anatole&&arnold&\cr
  \multispan{11}\vrulefill\cr
  ...
  &desdemona&&diane&\cr
  \multispan5\vrulefill\cr}
```

abel	basil	cecilia	desdemona
agnes	bernard	christopher	diane
amanda	bertrand	cuthbert	
anatole	brigitte		
arnold			

10
Tabbing

The previous chapter discussed the very general command for making tables. There is another facility in plain TeX that is easier to use in some applications: it is inspired on the idea of setting tabs on a typewriter. The tabs mark certain horizontal positions on the page, and writing texts starting at those positions is very easy.

In TeX, a line that should obey tabs starts with the \+ command, and the tabs themselves are represented by ampersands &. The end of the line is marked by \cr:

$$\texttt{\textbackslash+ ... \& ... \& ... \& ... \textbackslash cr}$$

10.1 Setting tabs

Room	8 to 10am	10am to noon
C8	Foata	Désarménien
C9	Schiffmann	Martinet
C10	Colloquium: Prof. Victor Ostromoukhov	

To get the alignment above, we typed

```
\+ \kern .8in  & \kern 1.2in   &                   \cr
\+ Room        & 8 to 10am     &10am to noon     \cr
\+ C8          &Foata          &D\'esarm\'enien \cr
\+ C9          & Schiffmann    &Martinet         \cr
\+ C10         & Colloquium: Prof.~Victor Ostromoukhov\cr
```

When TeX sees a \+, it starts typesetting a horizontal box. As it encounters a tab, say the third from the \+, it checks to see if it knows how far it the third column should be from the left margin. If it does, it skips to that position. If not, it sets the tab for the position that it's currently in, based on all the material since the \+. The first line in the code above places two tabs: the first .8 in form the left margin, and the second 1.2 in from the preceding one. The remaining lines then use the positions that were set on the first.

TEX ignores spaces after the \+ and after the &, so Foata and Schiffmann are aligned on the output, even though they're not in the listing above. But the spaces after an entry and before the next & are not ignored, although here they make no difference.

A tab entry can spill over to the next column, as on the last line of the table shown. On a typewriter, this would cause the tab that marks the next column to be subsequently skipped. But TEX backtracks to the beginning of the current column before advancing to the next: if we had ...Ostromoukhov & Reception, TEX would still align Reception with Martinet, overlapping with the previous entry.

Tabs don't have to be set all at once: whenever TEX finds a & to the right of all existing tabs, it sets a new one. For this Pascal triangle, we set tabs as needed, one per line. Notice the \quad after each entry that fixes the next tab's position: without it, there would be no spacing between the columns.

```
$$\vbox{
\+ 1\cr                          1
\+ 1\quad &1 \cr                 1  1
\+ 1 &2\quad &1 \cr              1  2  1
\+ 1 &3 &3\quad &1 \cr          1  3  3  1
\+ 1 &4 &6 &4\quad &1\cr        1  4  6  4  1
}$$
```

10.2 Centering

The fact that tabbed lines are composed in individual \hboxes makes it very easy to stack them together and center the whole assemblage:

```
$$\vbox{
    \+ ............. \cr
    ..................
    \+ ............. \cr
}$$
```

As with \halign (section 9.3), it's worth following a discipline when typing this type of code. Start with the outside layers—the dollar signs and the \vbox—then fill them in with the tabbed lines. This way you're less likely to forget to close the braces and the $$.

Tabs provide an alternative solution to the common problem of setting several short lines of text in a box, in such a way that the box comes out only as wide as the longest line inside (remember that if you say \vbox{line 1\par line 2}, you get a box of width `\hsize`). We've already seen how `\hbox` can be used in this case; `\+` gives an equivalent construction:

```
\vbox{                          \vbox{
    \hbox{line 1}                   \+ line 1\cr
    \hbox{line 2}                   \+ line 2\cr
    ...........                     ...........
}                               }
```

10.3 Choosing column widths

It's not always clear how to choose the widths of the columns when setting tabs. How do you know, except by trial and error, how many inches you should leave for the text in the first column?

If you know in advance what the widest entry in each column will be, there is an easy solution. Make a sample line containing the widest entries, but precede it with `\settabs` command. This will set the tabs as a normal `\+` line would, but the sample line will not appear on the output. Don't forget to include some spacing in the sample line entries, so the columns won't touch each other:

```
\settabs
\+ Room\qquad & Schiffmann\qquad &D\'esarm\'enien\cr
\+ Room       & 8 to 10am        &10am to noon\cr
...
```

10.4 Equally spaced tabs

The `\columns` command works with `\settabs` as an abbreviation for a common case: dividing up a page into columns of the same width. For concreteness, assume that the `\hsize` is 5 in; then `\settabs 5\columns` has the effect of `\settabs\kern 1in&\kern 1in&\kern 1in&\kern 1in&\cr`.

```
\settabs 5\columns
\+ Room & 8 to 10am    &10am to noon\cr
...
```

The syntax is somewhat inconsistent: `\columns` doesn't take `\+` or `\cr`, unlike the sample line construction of the previous section.

10.5 Clearing tabs

Once you have set a tab, it will ordinarily remain in effect until the next `\settabs` command. This means that you can have your tables obey the same alignment, even if there is normal text between them, or they are on different pages.

But that's not always what you want; it may be better to have different tables align differently. One solution is to wrap each table in braces: tabs defined inside a group disappear at the end of the group, and the ones in effect before the group was entered, if any, are restored. There's no way around that; it's an error to say `\global\settabs`.

There's also a `\cleartabs` command, which does what its name says. If you don't want to risk getting confused with previously set tabs, you should systematically start your tables with either `\cleartabs` or `\settabs`. If no tabs were set, no harm done—better safe than sorry.

You can also use `\cleartabs` inside a line of the form `\+...\cr`, in which case it clears only the tabs to the right of the column it's in. Any ampersands past that column will then set new tabs.

```
\cleartabs
\+ \quad& aaa\quad& bbb\quad& \cr
\+       &          &         & ccc\quad& ddd\quad\cr
\+       & AAA      & BBB     & CCC     & DDD      \cr
\+       & xxxxxxx\quad\cleartabs& yyyyy\quad & zzzzz\cr
\+       & XXX                  & YYY        & ZZZ  \cr
```

In the code above, the first `\cleartabs` cleans the slate, and three tabs are set after the first `\+`. On the next line one tab is set, and on the next none: the same ones are used. On the next line, `\cleartabs` leaves alone the first tab and clears the others; but two new tabs are immediately reset:

```
aaa   bbb
          ccc   ddd
AAA BBB CCC DDD
xxxxxxx  yyyyy  zzzzz
XXX      YYY    ZZZ
```

Often, `\cleartabs` is immediately followed by an ampersand to set a new tab. If your table doesn't contain the character &, we suggest you redefine locally the control sequence `\&`, by saying

$$\def\&{\cleartabs \&}$$

Put this definition inside the `\vbox` that contains your table (if you're centering it), or start a group just for this purpose. This way you won't forget it later on. See an example of use in section 10.8.

10.6 Tabs and rules

Horizontal rules

To obtain horizontal rules, we use `\hrule` between the end of a row (`\cr`) and the beginning of the next (`\+`):

Room	8 to 10am	10am to noon
C8	Foata	Désarménien
C9	Schiffmann	Martinet
C10	Colloquium: Prof. Victor Ostromoukhov	

```
$$\vbox{\def\strut{\vrule height 11pt depth 5pt width 0pt}
  \settabs
  \+ Room\qquad  & Schiffmann\qquad &D\'esarm\'enien\cr
  \+ \strut Room &8 to 10am  &10am to noon\cr
  \hrule
  \+ \strut C8   &Foata       &D\'esarm\'enien\cr
  \hrule
  \+ \strut C9   &Schiffmann &Martinet\cr
  \hrule
  \+ \strut C10  &Colloquium: Prof.~Victor Ostromoukhov\cr
}$$
```

The \strut s are used to keep the rules from sticking to the text. They're defined in plain T_EX to have height 8.5 pt and depth 3.5 pt, but we increase that here for better visual effect. (See also section 9.9.)

When you use rules, it is essential to wrap everything in a box. Just for fun, let's see what happens if we eliminate the surrounding \vbox and the \strut s:

Room	8 to 10am	10am to noon
C8	Foata	Désarménien
C9	Schiffmann	Martinet
C10	Colloquium: Prof. Victor Ostromoukhov	

Not a very pleasant result! The rules are much longer than the table; the reason is that they don't have a surrounding box, so they extend to the width of the page.

To obtain a horizontal rule across one column only, you can use, according to taste:

- \hrulefill, if the column width is already set (since a spring cannot make room for itself); or

- \vrule height .4pt width ... (remember that you can't use \hrule in horizontal mode). With this solution you kill two birds with one stone: you define the column width and set a new tab.

Here we use the second solution, setting tabs as we go along:

```
$$\vbox{\offinterlineskip\cleartabs
  \def\hr{\vrule height .4pt width 2em}
  \def\vr{\vrule height12pt depth 5pt}
  \def\cc#1{\hfill#1\hfill}
  \+ \hr&\cr
  \+ \vr\cc{1}&\vr\cr
  \+ \hr&\hr&\cr
  \+ \vr\cc{1}&\vr\cc{1}&\vr\cr
  \+ \hr&\hr&\hr&\cr
  ...
  \hrule
}$$
```

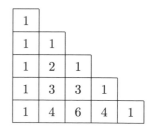

Vertical rules

The vertical rules \vr in the previous example, while part of each row, piece together to form an unbroken rule. To achieve this effect it is essential to eliminate the automatic interline spacing, using \offinterlineskip, as we did in chapter 9. But in contrast with the situation there (section 9.10), all the vertical rules must be give an explicit height and depth, because they are set in separate boxes.

Room	8 to 10am	10am to noon
C8	Foata	Désarménien
C9	Schiffmann	Martinet
C10	Colloquium: Prof. Victor Ostromoukhov	

Here the sample line after \settabs sets three tabs, rather than two, because the last \vr must be in a column by itself. Unfortunately, the width of the last column has to be set manually to 1.8 inches, because it depends on the long Colloquium entry, which starts in the previous column.

```
$$\vbox{\offinterlineskip
 \def\vr{\vrule height 12pt depth 5pt} \def\vrq{\vr\quad}
 \settabs
 \+\vr\quad Room\quad&\vr\quad Schiffmann\quad&\kern 1.8in&\cr
 \+\strut\quad Room&\quad 8 to 10am      &\quad 10am to noon\cr
 \+\vrq C8 &\vrq Foata       &\vrq D\'esarm\'enien      &\vr\cr
 \+\vrq C9 &\vrq Schiffmann  &\vrq Martinet             &\vr\cr
 \+\vrq C10&\vrq Colloquium: Prof.~Victor Ostromoukhov&&\vr\cr
}$$
```

Adding \hrule s between the lines of this table, after each \cr, we get

Room	8 to 10am	10am to noon
C8	Foata	Désarménien
C9	Schiffmann	Martinet
C10	Colloquium: Prof. Victor Ostromoukhov	

10.7 Tabs and springs

You can move around the material in a column by using springs \hfill. They must be strong springs: \hfil won't do, because plain TeX already puts an \hfil in each box to take up the slack.

Let's redefine \vrq in the previous example to mean \vr\hfill:

Room	8 to 10am	10am to noon
C8	Foata	Désarménien
C9	Schiffmann	Martinet
C10	Colloquium: Prof. Victor Ostromoukhov	

The result isn't too pleasant, but it has the virtue of showing that spaces at the end of an entry are not ignored: compare the C9 and C10 boxes.

It also seems that `\hfill` didn't work with the entry that spans two columns. This is a much more subtle problem: since the `\+` command, unlike `\halign`, sets one entry at a time, there is no way TEX can place the second entry flush right in the third column. To right-justify the long entry we must move it to the third column, and use `\hidewidth` to let it spill into the previous column (cf. section 9.5):

```
\+\tvq C10&\tv&\hidewidth Colloquium ... Ostromoukhov&\tv\cr
```

Also, `\hfill` won't work in the rightmost entry of any column: there must be a `&` to fix the right boundary of the box.

10.8 Typesetting code

Typesetting a program in a structured language is an interesting application of tabs, because of the use of indentation in displaying the program structure. The amount of indentation changes as we go along the program:

$$
\begin{aligned}
&\textbf{var } x\colon \textbf{array}[0..10] \textbf{ of integer};\\
&\qquad i\colon \textbf{integer};\\
&i := 1;\\
&\textbf{while } x[i] <> 0 \textbf{ and } i < 9 \textbf{ do begin}\\
&\qquad x[i] := x[i-1];\ i := i+1;\\
&\qquad \textbf{if } sum = 0 \textbf{ then } sum := sum + 1\\
&\qquad \textbf{else begin}\\
&\qquad\qquad sum := sum - delta;\\
&\qquad\qquad x[i] := sum;\\
&\qquad\quad \textbf{end}\\
&\quad \textbf{end}
\end{aligned}
$$

Here's how this program was typeset:

```
$$\vbox{\def\&{\cleartabs &}
  \def\<#1>{\hbox{\bf#1}}  \def\[#1]{\hbox{\it#1\/}}

  \+\<var>   \&$x$: \<array>[0..10] \<of> \<integer>; \cr
  \+         &$i$: \<integer>; \cr
  \+$i:=1$; \cr
  \+\<while> \&$x[i]<>0$ \<and> $i<9$ \<do> \<begin>\cr
  \+         &$x[i]:=x[i-1]$; $i:=i+1$;\cr
  \+         &\<if> $\[sum]=0$ \<then> $\[sum]:=\[sum]+1$\cr
  \+         &\<else> &\<begin>\cr
  \+         &        &$\[sum]:=\[sum]-\[delta]$; \cr
  \+         &        &$x[i]:=\[sum]$; \cr
  \+         &        &\<end>\cr
  \+         &\<end>\cr
}$$
```

As explained in section 10.5, we use the abbreviation \& to clear tabs and immediately set another one. Notice the unusual definition of the control sequences \[and \<, which typeset their arguments in italics and boldface, as is conventional for variable names and reserved words, respectively. Their arguments are not surrounded by braces, but rather by the control sequence itself and a delimiter,] or >. For more details, see section 12.4.

10.9 Tabs and alignments: a comparison

The two ways to typeset tables—using \halign or tabs—have much in common. But there are also substantial differences: in general, \halign is more powerful and can handle a greater variety of tasks, while tabbing is quicker to get started with. Experienced users tend to use \halign about 90% of the time, reserving \+ for special situations, like tables extending over several pages and program code.

Here's a summary of the features of both facilities:

Feature	Tabbing	\halign
Is the same alignment saved from one table to another?	Yes	No
Can column positions be redefined?	Yes	No
Can arbitrarily long alignments be handled? (\halign memorizes the whole alignment before setting it.)	Yes	No
Can tables be broken across pages?	Yes	Yes
Can columns be centered or right-aligned? (With tabbing, springs must be added to all entries.)	Yes	Yes
Can entries span several columns? (With tabbing, this interferes with springs.)	Yes	Yes
Can horizontal and vertical rules be set? (With tabbing, rule specifications must be repeated.)	Yes	Yes
Are column widths computed automatically?	Sometimes	Yes
How easy is it to handle exceptions?	Hard	Easy
Can common features be "factored out?"	No	Yes
Can columns be separated automatically?	No	Yes
Can the alignment width be predetermined?	No	Yes

11
Typesetting mathematics

11.1 Generalities

To typeset mathematical symbols in the middle of text, surround them with single dollar signs: `$...$`. To obtain

$$\text{for every real } x, \text{ we have } \sin(2x) = 2\sin x \cos x$$

we went into math mode and out again twice:

```
for every real $x$, we have $\sin(2x)=2\sin x \cos x$.
```

If you want to display a formula, whether for emphasis or because the formula is too long or too tall to fit comfortably on a line, use two dollar signs. The splendid spectral sequence

$$E_2^{p,q} = H^p\bigl(H^q(X; \mathcal{A}^*(X;L) \otimes \mathcal{B})\bigr) \implies H^{p+q}(X; \mathcal{B})$$

was typeset with `^_`

```
$$
E_2^{p,q}=H^p\bigl(H^q(X;{\cal A}^*(X;L)\otimes
{\cal B})\bigr)\ \Longrightarrow\  H^{p+q}(X;{\cal B})
$$
```

As you can see, a displayed formula is centered and surrounded with a bit of spacing above and below. This spacing is given by the variables `\abovedisplayskip` and `\belowdisplayskip`, which you can adjust to your taste; plain TEX sets them as follows:

```
\abovedisplayskip=12pt plus 3pt minus 9pt
\belowdisplayskip=12pt plus 3pt minus 9pt
```

There are also `\abovedisplayshortskip` and `\belowdisplayshortskip`, used when the line preceding the formula is so short that the regular skips would leave a visual "hole." Their values in plain TeX are

```
\abovedisplayshortskip=0pt plus 3pt
\belowdisplayshortskip=7pt plus 3pt minus 4pt
```

It is by no means necessary to have the `$$` on a line by itself in your source file, though it does help see what's going on. But don't leave a blank line before the first `$$` or after the second, unless you really want to start a new paragraph (in which case you'll get more spacing, and the next line will be indented). And don't even think of leaving a blank line inside the formula: TeX cannot start a new paragraph in math mode, so it assumes that something went wrong and ends the formula.

Pairs of single dollar signs `$...$` and pairs of double dollar signs `$$...$$` delimit groups. TeX doesn't allow you to nest these groups directly: if you're in math mode and type a `$`, you get out of it. This is sometimes unfortunate, but can be circumvented: see the use of `\ifmmode` in section 12.9.

The correct use of spacing in mathematical formulas is a hallmark of a good typographer or typesetting system, and is governed by fairly complex traditional rules. TeX avoids burdening the user with this question by managing itself all the spacing. As a consequence, it also ignores all spaces and carriage returns in math mode: `$\int_a^bf(x)dx$` and `$\int _ a ^ b f ( x ) d x$` have exactly the same effect. (Of course, a space inside `\int` would not be allowed.) The only uses of spaces in a formula are to mark the end of control sequences and to make the source file more intelligible. In particular, you should make liberal use of carriage returns.

Although sophisticated, TeX's spacing mechanism is not perfect. There will be times when you'll want to change the amount of space TeX puts between the various elements of a formula. We'll soon see commands that make it easy to do so.

11.2 Math symbols

TeX divides the symbols and characters accessible in math mode into eight classes (see *The TeXbook*, page 154):

0. ordinary characters	4. opening delimiters
1. large operators	5. closing delimiters
2. binary operators	6. punctuation
3. relational operators	7. variable-family characters

This classification is what enables TeX to manage spacing in the sophisticated way we've mentioned (*The TeXbook*, page 170). Class 7 contains all digits and all lowercase and uppercase letters; they're called variable-family characters because they change font according to the current `\fam`. Class 6 contains the comma, the semicolon, and a special colon obtain by typing `\colon`. We won't discuss those two classes any further.

Ordinary characters

Ordinary characters comprise the decimal point, Greek letters, calligraphic capitals (discussed in section 11.3) and certain math symbols:

α	\alpha	ι	\iota	ϱ	\varrho
β	\beta	κ	\kappa	σ	\sigma
γ	\gamma	λ	\lambda	ς	\varsigma
δ	\delta	μ	\mu	τ	\tau
ϵ	\epsilon	ν	\nu	υ	\upsilon
ε	\varepsilon	ξ	\xi	ϕ	\phi
ζ	\zeta	o	o	φ	\varphi
η	\eta	π	\pi	χ	\chi
θ	\theta	ϖ	\varpi	ψ	\psi
ϑ	\vartheta	ρ	\rho	ω	\omega

Lowercase Greek letters

There are only eleven uppercase Greek letters that don't look like some letter from the Latin alphabet:

Γ	\Gamma	Ξ	\Xi	Φ	\Phi
Δ	\Delta	Π	\Pi	Ψ	\Psi
Θ	\Theta	Σ	\Sigma	Ω	\Omega
Λ	\Lambda	Υ	\Upsilon		

Uppercase Greek letters

The other assorted mathematical symbols are:

$\aleph$	\aleph	$\prime$	\prime	$\forall$	\forall
$\hbar$	\hbar	$\emptyset$	\emptyset	$\exists$	\exists
$\imath$	\imath	∇	\nabla	$\neg$	\neg *or* \lnot
$\jmath$	\jmath	$\surd$	\surd	$\flat$	\flat
ℓ	\ell	$\top$	\top	$\natural$	\natural
$\wp$	\wp	$\bot$	\bot	$\sharp$	\sharp
$\Re$	\Re	$\|$	| *or* \Vert	$\clubsuit$	\clubsuit
$\Im$	\Im	$\angle$	\angle	$\diamondsuit$	\diamondsuit
∂	\partial	$\triangle$	\triangle	$\heartsuit$	\heartsuit
∞	\infty	$\backslash$	\backslash	$\spadesuit$	\spadesuit

Other ordinary characters

You can make any math expression or symbol into an ordinary symbol, for the purposes of spacing, by surrounding it with braces: $a+b$ gives $a + b$, but $a{+}b$ gives $a+b$.

Large operators

Large operators come in two sizes: very large, for display math, and not-so-large, for regular math. (The distinction is actually according to the style: see section 11.7.)

Large operators

Binary operators

Several binary operators have large counterparts, like $\cup$ and $\bigcup$. Mathematicians will have no trouble making the distinction, but if you're a non-mathematician and run across one of these the following rule of thumb may be helpful: the binary form is used between two letters or expressions ($A \cup B$), while the large form is used when there is only one expression, generally with subscripts, following the operator ($\bigcup A_i$).

Another possible source of confusion is \setminus, denoting set difference. It gives the same symbol as \backslash, but surrounds it with more spacing, as is usual with binary operators. See the Dictionary for examples.

To eliminate the spacing around a binary operator you can use the trick explained on the previous page: $u{\circ}v$ gives $u{\circ}v$, which some prefer over $u \circ v$.

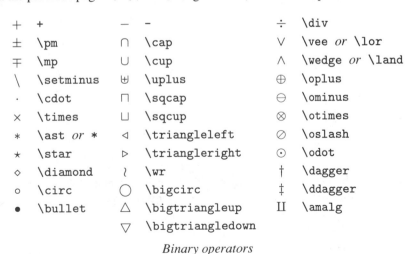

Binary operators

Relational operators

There are lots of them! The only thing to remark is that \mid and \parallel give the same symbols as | and \|, but put more space around them. See examples in the Dictionary.

<	\<	>	\>	=	=
≤	\leq *or* \le	≥	\geq *or* \ge	≡	\equiv
≺	\prec	≻	\succ	~	\sim
⪯	\preceq	⪰	\succeq	≃	\simeq
≪	\ll	≫	\gg	≍	\asymp
⊂	\subset	⊃	\supset	≈	\approx
⊆	\subseteq	⊇	\supseteq	≅	\cong
⊑	\sqsubseteq	⊒	\sqsupseteq	⋈	\bowtie
∈	\in	∋	\ni	∝	\propto
⊢	\vdash	⊣	\dashv	⊨	\models
⌣	\smile	|	\mid	≐	\doteq
⌢	\frown	‖	\parallel	⊥	\perp
		:	:		

Relational operators

Most of these relations can be negated by preceding their control sequences with \not. The resulting symbols are obtained by overstriking the original operator with a slash; since the width of the operators vary, some of them don't look exactly right. In particular, \not\in has a variant, written \notin, which looks somewhat better: the former gives ∉, the latter ∉.

≮	\not<	≯	\not>	≠	\not=, \neq, \ne
≰	\not\leq	≱	\not\geq	≢	\not\equiv
⊀	\not\prec	⊁	\not\succ	≁	\not\sim
⋠	\not\preceq	⋡	\not\succeq	≄	\not\simeq
⊄	\not\subset	⊅	\not\supset	≉	\not\approx
⊈	\not\subseteq	⊉	\not\supseteq	≇	\not\cong
⋢	\not\sqsubseteq	⋣	\not\sqsupseteq	∉	\notin

Negations

Arrows are a special type of relations. Most horizontal arrows come in two sizes, the longer of which starts with long. Vertical arrows also grow, but with different prefixes: see section 11.13. In this respect they work like delimiters (see the next subsection).

The \iff command gives the same symbol as \Longleftrightarrow, but places extra space on both sides.

←	\leftarrow, \gets	⟵	\longleftarrow	↑	\uparrow
→	\rightarrow, \to	⟶	\longrightarrow	↓	\downarrow
↔	\leftrightarrow	⟷	\longleftrightarrow	↕	\updownarrow
⇐	\Leftarrow	⟸	\Longleftarrow	⇑	\Uparrow
⇒	\Rightarrow	⟹	\Longrightarrow	⇓	\Downarrow
⇔	\Leftrightarrow	⟺	\Longleftrightarrow	⇕	\Updownarrow
↦	\mapsto	⟼	\longmapsto	↗	\nearrow
↩	\hookleftarrow	↪	\hookrightarrow	↘	\searrow
↼	\leftharpoonup	⇀	\rightharpoonup	↙	\swarrow
↽	\leftharpoondown	⇁	\rightharpoondown	↖	\nwarrow
		⇌	\rightleftharpoons		

Arrows

Left and right delimiters

Parentheses, brackets, and other symbols that come in pairs are collectively known as delimiters, and TeX makes it easy to get them in different sizes (section 11.13). Here are the basic control sequences to obtain them. Brackets [ ] and braces { } have alternative names, since unfortunately they are not available on some keyboards.

(	(	[	[*or* \lbrack	⌊	\lfloor
)	)	]	] *or* \rbrack	⌋	\rfloor
⟨	\langle	{	\{*or* \lbrace	⌈	\lceil
⟩	\langle	}	\}*or* \rbrace	⌉	\rceil

Left and right delimiters

11.3 Fonts in math mode

Generally, letters in mathematical formulas are typeset in italics, to stand out better from the surrounding text. TeX's main font for math is called math italic, or `cmmi`, with slightly wider characters than text italic (compare a and *a*).

As we discussed in section 4.7, the font change commands `\rm`, `\bf`, `\it`, `\sl` and `\tt` work in math mode, but with the last three you can't use subscripts or superscripts.

`$Pqr+{\rm Xyz}+{\bf Uvw}+{\it Xyz}$` ..... $Pqr + \mathrm{Xyz} + \mathbf{Uvw} + Xyz$

But if you say `\hbox{...}` in math mode, you're in horizontal mode inside the braces, and you revert to the font situation before the dollar signs:

`AAA {\bf BBB $(xy+\hbox{xy})$}` .... AAA **BBB** $(xy + \mathbf{xy})$

`CCC $(uvw+\hbox{uvw})$` .... CCC $(uvw + \mathrm{uvw})$

Here are some other miscellaneous font commands. As usual, they affect all letters till the end of the group in which they are issued; so it's best to confine their action using groups, or you'll get wierd results (see the Dictionary under `\cal`).

- \mit can be used for italic Greek capitals (see section 11.2 for the letter names): ${\mit\Gamma}$ gives $\mathit{\Gamma}$, to be compared with the usual Γ. On digits it has the same effect as \oldstyle below.

- \cal is used for calligraphic letters: ${\cal ABC}$ gives $\mathcal{ABC}$. Only capitals are available.

- \oldstyle produces old-fashioned digits: {\oldstyle 1234567890} will give 1234567890. You can use this command outside math mode: it is handy for bibliographies (see section 4.8).

Section 4.7 discusses in detail what you must do to define new fonts in math mode.

11.4 Subscripts and superscripts

Subscripts are obtained with the underscore character: x_k gives x_k. Ordinarily, only the character following the _ is made into a subscript; if you want the subscript to consist of several characters you must group them:

$a_i+b_{i,j}$ .... $a_i + b_{i,j}$

Another case where braces are indispensable is in resolving an ambiguous double subscript: what does x_k_i mean? You must choose either ${x_k}_i$, which gives x_{k_i}, or x_{k_i}, which gives x_{k_i}. (But the former almost never occurs.)

Superscripts are very much like subscripts, and you get them a caret, also known as hat: e^x gives e^x. Here, too, watch out for ambiguous expressions like x^k^2; you must use braces around x^k (rare) or around k^2.

$N=2^{2^{2^{2^{2^2}}}}$ .... $N = 2^{2^{2^{2^{2^2}}}}$

Subscripts and superscripts can be combined, in either order. In general, mathematicians think of the subscript as more closely related to the main letter, so they type it first. But TEX will print exactly the same thing if you type the superscript first:

$x_{r,s}^{p+q}+y^{p+q+1}_{m^2}$ .... $x_{r,s}^{p+q} + y_{m^2}^{p+q+1}$

The empty group

It is possible to adorn the empty group with subscripts and superscripts. This has several uses, such as

- placing subscripts and superscripts before a letter: the isotope ${}^{238}_{92}\mathrm{U}$ can be coded ${}^{238}_{\hphantom{0}92}{\rm U}$ (for \hphantom, see 9.8);

- staggering subscripts and superscripts: if you're fond of relativity theory and its tensors, write $\Gamma_{i,j}{}^r{}_k$ for $\Gamma_{i,j}{}^r{}_k$;

- aligning subscripts and superscripts: if you dislike the fact that TEX normally shifts superscripts slightly (H_2^2), to account for the slant of italic letters, place an empty group after the letter, as in $H{}_2^2$, which gives $H{}_2^2$.

Primes

To conclude, there is what is probably the most common superscript in mathematics, the so-called prime. It is so common that it has a shorthand form `'` :

`$(u\cdot v)'=u'\cdot v +u\cdot v'$` .... $(u \cdot v)' = u' \cdot v + u \cdot v'$

The full form would be `$u^\prime$` , and so on. Plain TEX also arranges things so that `'` can be repeated, or combined with other superscripts, without the need for grouping: `$x''$` gives x'', and `$x'^2$` gives x'^2, without further ado.

11.5 Accents

Accents and math don't go well together: TEX will complain if you give it the commands of section 2.4 while in math mode. To typeset $H^*_{\text{étale}}(X; \mathcal{F})$, then, `$H_{\rm\'etale}^*(X;{\cal F})$` won't work. The solution is to typeset the word that requires the accent in an `\hbox` (this unfortunately requires explicitly choosing a smaller font for the subscript):

`$H_{\hbox{\sevenrm\'etale}}^*(X;{\cal F})$` .... $H^*_{\text{étale}}(X; \mathcal{F})$

Nonetheless, several of the accents in section 2.4 can be obtained in math mode, after all; they just have to be referred to by different names:

$\hat{a}$	`\hat a`	$\check{a}$	`\check a`	$\tilde{a}$	`\tilde a`
$\grave{a}$	`\grave a`	$\acute{a}$	`\acute a`	$\vec{a}$	`\vec a`
$\dot{a}$	`\dot a`	$\ddot{a}$	`\ddot a`	$\breve{a}$	`\breve a`
		$\bar{a}$	`\bar a`		

This table also includes other "accents" which are not available outside math mode, such as the very common `\vec x` , and the double dot:

`$\ddot x+x+ax^3=f(t)$` .... $\ddot{x} + x + ax^3 = f(t)$

All those new accents work in math mode only. They are meant for single letters; but sometimes you need an accent to cover more than one letter. With `\bar` and `\vec` there is no problem: they have counterparts that grow to match the expression underneath, which will be introduced in section 11.12.

There are also wide counterparts for `\hat` and `\tilde`, but they only grow to about three characters wide:

`$\widehat{xy}$, $\widetilde{xy}$` .... $\widehat{xy}, \widetilde{xy}$

`$\widehat{xyzw}$, $\widetilde{xyzw}$` .... $\widehat{xyzw}, \widetilde{xyzw}$

11.6 Spacing in math mode

We've seen that spaces in the input are ignored in math mode. How then can you create extra spacing in a formula? We'll discuss only the most common ways:

- \quad , \qquad and \ (backslash-space) work exactly as in horizontal mode (see section 5.1);
- \, is the math mode counterpart of \thinspace (section 5.5), and gives a thin space, which is about 1.5 pt wide in ten-point size;
- \! is the counterpart of \negthinspace and gives a negative space, that is, it brings things together by 1.5 pt.

These commands should be used judiciously: most of the time T_EX knows what it's doing, and you won't have to add space by hand at all. But this expression shows two cases where the automatic spacing can be improved on:

$$\int\int_{\cal D}f(u,v)dudv$$ . . . . $\displaystyle\int\!\int_{\mathcal{D}} f(u,v)dudv$

$$\int\!\!\!\!\int_{\cal D}f(u,v)\,du\,dv$$ . . . . $\displaystyle\iint_{\mathcal{D}} f(u,v)\,du\,dv$

A very common example of the use of \quad and \qquad is in separating explanations, conditions, etc., that accompany a displayed formula. The display

$$e^{ix} = \cos x + i \sin x \qquad \text{for every } x$$

was obtained with

$$e^{ix}=\cos x+i\sin x \qquad\hbox{for every x}$$

Notice the interplay between modes here: The easiest way to get straight text in a display is by using an \hbox , and inside the box you can start a new level of math mode if necessary. Saying ...\hbox{for every } x$$ also works, but seems less clear.

Apart from the commands discussed above, you can use \hskip and \kern at will in math mode, as well as the special commands \mskip and \mkern that we will discuss in the next section. Use all these commands sparingly: few things make the appearance of a document more unprofessional than an arbitrary sprinkling of inconsistent spaces.

11.7 The four styles

We have seen that all letters automatically come out smaller when they're used in subscripts, and also that some operators come out smaller in displayed equations than in the middle of a paragraph. The size of all characters is a function of the current *style*, which is a way of saying what part of an expression the character appears in. There are four styles, as follows:

When T_EX goes into display math mode (between double dollar signs), it typesets things in *display style*. In this style all characters have their full size, which, except for large operators like \int , is essentially the size of the surrounding text.

But when TₑX sees a subscript or superscript, it switches to *script style* to typeset it, and back again when it's done. In script style characters are somewhat smaller, like this: x.

And if, while already in script style, TₑX is required to set a subscript or superscript, it does it in *scriptscript style*, which is even smaller, like this: x. (But if TₑX sees a subscript while it's already in scriptscript style, it doesn't make it any smaller: there is no scriptscriptscript style.)

So in the expression

$$\int e^{-x^2}\, dx,$$

which comes from `$$\int e^{-x^2}\,dx,$$` the ^{2} is in scriptscript style, the $^{-x}$ in script style, and all the rest is in display style.

This takes care of three styles. How about the fourth? It is called *text style*, and it's the one that TₑX starts in when it enters normal math mode (between single dollar signs). In text style, too, letters come out the same size as the surrounding text, but large operators don't come out quite as large as in display style. If the integral of the previous paragraph had been set in normal math mode, it would look like this: $\int e^{-x^2}\, dx$.

For those who have read section 4.7: This explains why math fonts are kept in families. When the current family is `\fam1`, say, TₑX will fetch an 'a' from `\textfont1` if the current style is display or text; from `\scriptfont1` if the style is script, and from `\scriptscriptfont1` if the style is scriptscript. This explains why the size changes with the style: `\scriptfont1` is a smaller font than `\textfont1`, and so on.

The size of large operators is not the only difference between display and text styles. In text style, TₑX takes great pains to limit the height and depth of formulas, so they will not interfere with the spacing between lines; for this reason, whenever it has to stack things vertically, as in a fraction, it chooses to switch to script style. Consider the difference between the next two lines:

`$x+{y\over 1+x^2}$` .... $x + \frac{y}{1+x^2}$

`$$x+{y\over 1+x^2}$$` .... $x + \dfrac{y}{1 + x^2}$

In the first case, TₑX starts in text style, goes into script style for the numerator and denominator of the fraction, and into scriptscript style for the exponent '2'. In the second case, it starts in display style, goes into text style for the fraction's components, and into script style for the exponent.

You can change styles at any point in any formula, by saying `\displaystyle`, `\textstyle`, `\scriptstyle` or `\scriptscriptstyle`. The change will stay in effect until the end of the current group, with consequent changes in the style of fractions, subscripts, and so on.

For example, `$e^{x\over n}$` gives $e^{\frac{x}{n}}$: the exponent's style is script, so the style of the numerator x and denominator n is scriptscript. This seems to be one of the few cases where TEX's decisions are unfortunate, because the result is too small and the spacing wrong. The solution is to magnify both numerator and denominator a notch by setting them in script style:

`$e^{\scriptstyle x\over \scriptstyle n}$` . . . . $e^{\frac{x}{n}}$

You could also try `$e^{\textstyle{x\over n}}$`, doing the whole fraction in text style, instead of numerator and denominator individually in script style. The letters would come out the same size, but the spacing would be different. For another example, see section 11.9.

(If you haven't read section 4.7, read it now or skip to the next section.) TEX has a unit of length that depends on the current math font, much like an em depends on the current font. It's called a mu (for *mathematical unit*), and it's exactly 18 times smaller than the current math font's em, for some obscure reason known only to Knuth. Spaces measured in mus are obtained with the `\mskip` and `\mkern` commands, which are like `\hskip` and `\kern` but demand this particular unit. For example, `\,` is equivalent to `\mkern 3mu` (notice that `mu` doesn't have a backslash, unlike the greek letter μ). The current math font is a function of the current `\fam` and of the current style, so glue and kerns specified in mus scale correctly when the style changes. This is especially advantageous inside macros.

11.8 Function names

To improve readability, certain multi-letter abbreviations for function names are traditionally set in roman, and surrounded with a little bit of space: $\sin x$ versus $sinx$. TEX will do that automatically if you remember to precede the abbreviation with a backslash:

`$\sin^2 x+\cos^2 x=1$` . . . . $\sin^2 x + \cos^2 x = 1$

Here is a list of such control sequences provided by plain TEX:

`\arccos`	`\cos`	`\csc`	`\exp`	`\ker`	`\limsup`	`\min`	`\sinh`
`\arcsin`	`\cosh`	`\deg`	`\gcd`	`\lg`	`\ln`	`\Pr`	`\sup`
`\arctan`	`\cot`	`\det`	`\hom`	`\lim`	`\log`	`\sec`	`\tan`
`\arg`	`\coth`	`\dim`	`\inf`	`\liminf`	`\max`	`\sin`	`\tanh`

See also section 11.10, and the Dictionary under `\bmod` and `\pmod`.

11.9 Fractions

Fractions are made with the `\over` command: we've seen it a few times already. The action of `\over` extends over the smallest containing group, which may be the whole formula or something delimited with braces: `$$a+b\over a-b$$` and `$$a+{b\over a}-b$$` give

$$\frac{a+b}{a-b} \quad \text{and} \quad a+\frac{b}{a}-b.$$

We've seen that if TeX sees a fraction while in display style, it sets its components in text style; if it sees it in text style the components come out in script style; and if it sees it in script style the components come out in scriptscript style. Beyond that there is no change.

If you prefer to have $\dfrac{ab}{a+b}$ rather than $\frac{ab}{a+b}$ in a paragraph, change the style explicitly: `$\displaystyle {ab\over a+b}$`. (The braces are essential: if you write `$\displaystyle ab\over a+b$`, you get $\frac{ab}{a+b}$, because `\over` effectively makes each component of the fraction into a group.)

TeX makes the fraction bar the right length automatically, and centers the numerator and denominator with weak springs `\hfil`. Section 5.4 shows how you can set the shorter component flush right or flush left using `\hfill`.

Needless to say, a fraction in "slashed" form, like a/b, is coded `$a/b$`. The slash often offers a better alternative to complicated fractions in text: $a/(b+1)$ seems preferable to either $\frac{a}{b+1}$ or $\dfrac{a}{b+1}$. (Non-mathematicians take note: to preserve the meaning when setting a fraction in slashed form, you must wrap the numerator and denominator in parentheses, unless they consist of a single symbol!)

Continued fractions

These typographically fiendish fractions provide a good real-life example of the use of `\displaystyle`.

$$x = x_0 + \cfrac{1}{x_1 + \frac{1}{x_2 + \frac{1}{x_3}}} \qquad\qquad x = x_0 + \cfrac{1}{x_1 + \cfrac{1}{x_2 + \cfrac{1}{x_3}}}$$

The straightforward specification

```
$$x=x_0+{1\over x_1+{1\over x_2+{1\over x_3}}}$$
```

gives the hideous mess on the left. The size keeps decreasing: denominators are set, successively, in text style, script style and scriptscript style. The version on the right corrects that problem:

```
$$x=x_0+{1\over\displaystyle x_1+
{1\over\displaystyle x_2+{1\over\displaystyle x_3}}}$$
```

but the result is still not entirely satisfactory, because the '1's are too close to the bar. To "push up the ceiling," we resort to struts, invisible rules of predetermined height and depth. Plain TeX defines a strut especially for use in math formulas; it's called `\mathstrut`, and it is exactly the height and depth of parentheses in the usual math font (,). For more about struts, see sections 11.17 and 9.9.

```
$$x=x_0+{1\over\displaystyle x_1+
 {\mathstrut 1\over\displaystyle x_2+
  {\mathstrut 1\over\displaystyle x_3}
}}$$
```

$$x = x_0 + \cfrac{1}{x_1 + \cfrac{1}{x_2 + \cfrac{1}{x_3}}}$$

If this seems hopelessly complicated, don't worry: Rome wasn't made in a day!

Stacking other things

The \atop command works just like \over, but it doesn't draw the horizontal bar! This may seem pointless, until you hit something like

$$\sum_{\substack{i,j\in I \\ i\neq j}} x_i y_j$$

The code here was

```
$$\sum_{\textstyle {i,j\in I \atop i\ne j}}x_iy_j$$
```

Notice the request for the subscript to \sum to be set in text style. If it were set in script style, the default, its two lines would be in scriptscript style, and would look too small.

The \atop and \over commands are part of a larger family, which includes \above, \abovewithdelims, \atopwithdelims and \overwithdelims. These commands let you control the thickness of the fraction bar and automatically place parentheses or brackets around the fraction. Their use is rare, so we leave the details to the Dictionary. But we can mention here the macros \choose, \brack and \brace, which are made from \atopwithdelims and are used often enough (especially the first):

```
$2n \choose n$, $x \brack y$, $a \brace b$
```
..... $\displaystyle\binom{2n}{n}, \left[\begin{matrix}x\\y\end{matrix}\right], \left\{\begin{matrix}a\\b\end{matrix}\right\}$

11.10 Large operators and limits

The observant reader will have noticed in the previous section that the construction \sum_{...} sets the subscript under the summation sign, rather than in the usual place for subscripts. This happens in display style only: compare $\sum_{n=1}^{\infty} n^{-2}$ with $\sum_{n=1}^{\infty} n^{-2}$, both of which come from typing \sum_{n=1}^\infty n^{-2} after the appropriate style command.

Most of the large operators listed in section 11.2 behave in the same way; some, like \int, have their subscripts and superscripts always in their usual place. You can override this behavior with the \limits and \nolimits commands (the name comes from the fact that the expressions placed above and below, say, $\sum$ indicate the limits of summation). With \limits, subscripts and superscripts are placed above and below the operator, no matter what the style. With \nolimits they are always placed in their normal positions.

```
$$\sum\nolimits_{n=1}^\infty n^{-2}$$
```
.... $\sum\nolimits_{n=1}^{\infty} n^{-2}$

```
$$\int\limits_{-\pi}^{\pi}\cos^2x\,dx$$
```
.... $\int\limits_{-\pi}^{\pi} \cos^2 x \, dx$

The second construction helps save space if your formula is too long, while the first is useful with a tall formula.

Section 11.8 pointed out that many traditional abbreviations are predefined control sequences. Plain TEX defines these commands to work just like large operators; depending on their meaning, some take limits, others normal subscripts and superscripts. Normally you don't have to worry about which does which: they do what they're supposed to.

$$\max_{i\in I}x_i$$ .... $\max\limits_{i\in I} x_i$

$$\sin^2 x+\cos^2 x=1$$ .... $\sin^2 x + \cos^2 x = 1$

The mechanism by which these new "operators" are defined is the `\mathop` command. If you say

> `\def\limproj{\mathop{\rm lim\,proj}}`

typing $$\limproj_{j\in J} X_j$$ will give

$$\lim_{j\in J} \mathrm{proj}\, X_j$$

By default, a new `\mathop` is like `\sum`: it has limits above and below in display style, and on the right otherwise. You can also define a `\mathop` with `\nolimits` or `\limits`:

> `\def\trace{\mathop{\rm trace}\nolimits}`

Now $$\trace^2 A$$ will produce $\mathrm{trace}^2 A$. And, if you're modifying an existing operator that has been defined with `\limits` or `\nolimits`, you can make it revert to the default behavior by tacking on `\displaylimits`.

You can even make up your own large operators: after the definition

> `\font\cmrXVII=cmr17 \def\mysum{\mathop{\hbox{\cmrXVII S}}}`

typing `\mysum_{i=0}^{i=n}a_i` results in $\mathrm{S}_{i=0}^{i=n} a_i$ in display style, and $\mathrm{S}_{i=0}^{i=n} a_i$ in text style.

Notice how we got S: the font change command `\cmrXVII` was issued inside an `\hbox`, that is, outside math mode. Otherwise it would have no effect, because it doesn't go through the `\fam` mechanism (section 4.7).

There is one problem with this definition: the S will come out the same size whether or not it is displayed, and even if it belongs to a subscript. As usual, there is a way to fix this: see the Dictionary under `\mathchoice`.

Large operators and the baseline

The `\mathop` command has an additional subtlety. Consider the following output, where `\rule` is defined as `\hbox to .2in{\hrulefill}`:

$S\rule\mathop{S}\limits_1^1$ .... $S\rule{}{}\mathop{S}\limits_1^1$

The 'S' inside `\mathop` is sunk a bit below the baseline, and its upper and lower limits don't align exactly. The same experiment with an abbreviation shows neither phenomenon:

$\rule\mathop{\rm sin}$ .... $\rule{}{}\mathop{\mathrm{sin}}\limits_1^1$

It turns out that if the argument to `\mathop` is a single character, the character is vertically centered about the "axis" (see the end of chapter 8), and its limits are staggered to account for the slant, if any. Anything else is left alone:

`$\rule\mathop{\hbox{S}}\limits_1^1$` .... $\underline{}\,S_1^{\,1}$

Exercise

Typeset the following Christmas tree ornament, first in a display and then in a paragraph:

$$6_7^5 \, M^{\,4}_{\,8} \, {}_1^3 2$$

One solution for the display is

`$${\scriptstyle6}{}_7^5\mathop{M}_8^4{}_1^3 {\scriptstyle2}$$`

To get the same result in a paragraph, start with `\displaystyle` or do the middle part with `\mathop{M}\limits`.

11.11 Radicals

To get the square root sign over an expression, type `\sqrt{...}`. The sign extends both horizontally and vertically as needed:

```
$$\pi=2\times{2\over\sqrt{2}}
    \times{2\over\sqrt{2+\sqrt{2}}}
    \times{2\over\sqrt{2+\sqrt{2+\sqrt{2}}}}\times\cdots$$
```

gives

$$\pi = 2 \times \frac{2}{\sqrt{2}} \times \frac{2}{\sqrt{2+\sqrt{2}}} \times \frac{2}{\sqrt{2+\sqrt{2+\sqrt{2}}}} \times \cdots$$

But there are subtle distinctions in spacing between display style (left) and text style (right):

$$\sqrt{2+\sqrt{2+\sqrt{2+\sqrt{2}}}} \qquad\qquad \sqrt{2+\sqrt{2+\sqrt{2+\sqrt{2}}}}$$

Normally, the square root sign extends up and down only far enough to enclose the material inside, so different letters get signs that don't necessarily align:

`$\sqrt a +\sqrt b+\sqrt{1+x_i^2}$` .... $\sqrt{a} + \sqrt{b} + \sqrt{1+x_i^2}$

If you don't like that, make all the expressions the same height and depth by using a `\mathstrut`:

`$\sqrt{\mathstrut a} + ...$` .... $\sqrt{a} + \sqrt{b} + \sqrt{1+x_i^2}$

Other radicals are made with the `\root` command. Its syntax is somewhat unusual: `$\root 3\of{1+x^2}$` gives the cube root $\sqrt[3]{1+x^2}$. The '3' is in scriptscript style; if that's too small for you, it's easy to change it (`\root...\of` delimits a group):

`$\root\scriptstyle 3\of{1+x^2}$` .... $\sqrt[3]{1+x^2}$

`$$\root\scriptstyle 3\of{1+x^2}$$` .... $\sqrt[3]{1+x^2}$

11.12 Horizontally extensible symbols

The square root sign is not the only symbol that grows to accomodate its argument. TEX has other extensible symbols: bars and arrows above and below an expression extend horizontally, and all sorts of delimiters like parentheses and brackets extend vertically.

To place a bar above an expression, use `\overline{...}`. Notice the difference between that and the `\bar` command of section 11.4, which doesn't grow:

`$\overline{z+z'}=\bar z+\bar z'$` .... $\overline{z+z'} = \bar z + \bar z'$

You can also `\underline` an expression:

`$\underline{a+b+\cdots+y+z}$` .... $\underline{a+b+\cdots+y+z}$

Arrows are obtained with `\overrightarrow` and `\overleftarrow`. Again, there is a great difference between these and `\vec`, which is meant for a single letter:

`$\overrightarrow{AB}+\vec{BC}$` .... $\overrightarrow{AB} + \vec{BC}$

Unlike `\bar` and `\vec`, these extensible bars and arrows don't take into account the slant of italic letters. For better results, you can introduce a manual correction with a definition like

        `\def\ora#1{\overrightarrow{\mkern-2mu#1\mkern 2mu}}`

`$\overrightarrow{AB}$`, `$\ora{AB}$` .... $\overrightarrow{AB}, \overrightarrow{AB}$

To place braces above a formula, there is `\overbrace`, which takes a label as a superscript:

`\overbrace{x+\cdots+x}^{n\ \rm times}$` .... $nx = \overbrace{x+\cdots+x}^{n \text{ times}}$

If you place braces above two groups of different heights, you should uniformize their heights with `\mathstrut` (compare section 11.11 and the continued fraction example in 11.9):

$$2^{\ell_2}\ldots n^{\ell_n} = \overbrace{2\times\cdots\times 2}^{\ell_2}\times\cdots\times\overbrace{n\times\cdots\times n}^{\ell_n}$$

```
$$
2^{\ell_2}\ldots n^{\ell_n}=
\overbrace{\mathstrut 2\times\cdots\times 2}^{\ell_2}
\times\cdots\times
\overbrace{\mathstrut n\times\cdots\times n}^{\ell_n}
$$
```

There is also \underbrace , the opposite of \overbrace :

$$x^2 + x + 1 = \underbrace{\left(x + \frac{1}{2}\right)^2}_{>0} + \frac{3}{4} > 0$$

```
$$x^2+x+1=\underbrace{(x+{1\over 2})^2}_{>0}+{3\over 4}>0$$
```

All these symbols are made of elementary pieces combined together in predefined ways. You can also combine the same and other pieces in custom-made patterns: see \joinrel in section 11.15 and \relbar , \Relbar and \rhook in the Dictionary.

11.13 Vertically extensible symbols

All the delimiters listed in section 11.2, all vertical arrows and bars (which can also be used as delimiters), and some other characters like the slash /, can be made to grow. Up to a certain point, the growth is accomplished by a choice of existing characters; after they run out, T_EX can piece together special parts to make indefinitely large composites. There are two ways to pick a large delimiter: using \big and its friends you make the choice yourself, while using \left and \right you leave the choice to T_EX, on the basis of the contents. We start with \big .

To get a set of parentheses that is slightly bigger than usual, but still fits comfortably on a normal line of text, you type \bigl(...\bigr) . This is often employed to improve the intelligibility of formulas with a nested structure:

```
$f(x+g(y))$, $f\bigl(x+g(y)\bigr)$
```
.... $f(x + g(y)), \; f\bigl(x + g(y)\bigr)$

The 'l' in \bigl and the 'r' in \bigr are responsible for inserting a bit of space next to the delimiter, on the appropriate side. This is especially important in the case of openers and closers that look the same:

```
$\bigl||x|+|y|\bigr|$
```
.... $\bigl||x| + |y|\bigr|$

But if you want a vertical bar that is not part of an open-close pair—generally such a bar represents a relation—you should use \bigm ; the 'm' stands for middle, and causes T_EX to put spacing on both sides:

```
$\bigl\{x+y\bigm|x\in X, y\in Y\bigr\}$
```
... $\bigl\{x + y \;\big|\; x \in X, y \in Y\bigr\}$

The same formula without \bigm , $\{x + y | x \in X, y \in Y\}$, would definitely look wrong to a mathematician.

Finally, \big by itself is used mostly with the slash which, although mathematically a binary operator, traditionally doesn't get any spacing around it:

```
$\bigl(f(t)+1\bigr)\big/\bigl(tg(t)\bigr)$
```
..... $\bigl(f(t) + 1\bigr)\big/\bigl(tg(t)\bigr)$

Instead of \big, you can also say \Big, \bigg and \Bigg, all of which have left, right and middle variants and yield increasingly larger characters:

$$\left(\left(\left(\left((\ )\right)\right)\right)\right)$$

> \Biggl(\biggl(\Bigl(\bigl(() \bigr)\Bigr)\biggr)\Biggr)

It's perfectly legal to have a \bigl delimiter without a matching \bigr. In contrast, \left and \right work in pairs: if you say \left(...\right), TEX treats as a group the expression represented by ..., and after typesetting it, it flanks it with the smallest set of parentheses that is big enough to enclose the whole expression:

$$\verb|$$\left({d^2\over dx^2}+a\right)f=0$$| \ldots\ldots\ldots\ldots \left(\frac{d^2}{dx^2}+a\right)f=0$$

$$\verb|$$\left|{a\over x^2-y^2}\right|$$| \ldots\ldots\ldots\ldots \left|\frac{a}{x^2-y^2}\right|$$

Unlike the \big series, \left and \right are capable of constructing arbitrarily large delimiters, such as the parentheses that go around large matrices (see section 11.21). Angle brackets and slashes are exceptions: they can't be made from separate pieces, so if you ask for something too big you'll be disappointed.

Since \left and \right work as a group, things like \left(...&...\right) and \left(...{...\right)...} are illegal. Also, if you have several \lefts and \rights in the same formula, they must be properly nested, like all groups. But the corresponding delimiters don't have to match; it's OK to have \left(pair up with \right]. You can even pair something up with \right., where the period stands for an "invisible delimiter," needed just to keep TEX's matchmaker happy. For instance, the formula

$$f(x) = \begin{cases} x & \text{for } x < 1 \\ 2-x & \text{for } x \geq 1 \end{cases}$$

came from

> $$f(x)= \left\{x \hfill\hbox{for $x<1$}$$
> \atop 2-x \quad \hbox{for $x\ge 1$}\right.$$

This particular setup is so common that plain TEX provides a macro to deal with it more easily: we would have gotten the same result by typing

> f(x)=\cases{x   & for $x<1$ \cr
>   2-x & for $x\ge1$\cr}

This macro, in effect, sets up an alignment, which can have any number of rows terminated by \cr. On each row, everything to the left of the & is read in math mode, and set flush left against the left braces; but everything to the right of the & is read in horizontal mode, since it normally involves some text.

It would be nice not to have to worry about a multiplicity of \bigs, and always use \left and \right. But TEX's recipe for choosing delimiter sizes isn't perfect, and in some cases it has to be overridden. One fairly common situation is when the expression contains a large operator with limits: a human typesetter would let the limits hang out a bit, so as to use slightly less huge delimiters, but TEX doesn't know that. Compare:

$$\left(\sum_{n=1}^{\infty}\frac{1}{n^2}\right) \qquad\qquad \biggl(\sum_{n=1}^{\infty}\frac{1}{n^2}\biggr)$$

\left(...\right) \qquad\qquad \biggl(...\biggr)

Another case was shown above: $\big\|\,|x|+|y|\,\big\|$. If you type `\left...\right` instead of `\bigl...\bigr` here, you get $\|\,|x|+|y|\,\|$, which is awfully confusing.

Occasionally, too, you must adjust the spacing because of unfortunate coincidences in shape:

$$\prod_{k\ge 0}\frac{1}{(1-q^k z)} = \sum_{n\ge 0} z^n \bigg/ \prod_{1\le k\le n}(1-q^k)$$

```
\prod_{k\ge 0}{1\over (1-q^kz)}=\sum_{n\ge 0}z^n
\biggm/\!\!\prod_{1\le k\le n}(1-q^k)
```

All the extensible symbols discussed in this section are vertically centered about an imaginary axis that runs some 2.5 pt above the baseline (see the end of chapter 8). This means that they always grow up and down by equal amounts, so if you have a formula that is very tall but not deep, there will be a lot of white space underneath it, and the delimiters will be too big. Fortunately, this doesn't happen very often, because large operators, fractions, and the like also tend to be distributed roughly symmetrically with respect to the axis. But if you try to make a vertical arrow half an inch long, say, by writing something like

```
\left\uparrow\vbox to .5in{}\right.
```

you may be surprised: the arrow will extend almost as far down as it does up, and its total length will be almost one inch (more exactly, one inch minus twice the axis height). You must instead use a centered box of height plus depth equal to half an inch, which you can obtain by using `\vcenter`:

```
\left\uparrow\vcenter to .5in{}\right.
```

11.14 Stacking up symbols

Mathematicians are fond of creating new symbols, and one time-honored way of doing so is by adding bells and whistles to old ones. Plain TEX caters to that tradition in several ways, and we devote this section and the next two to some of them.

The `\buildrel` macro adorns relations and arrows—and just about anything else—by writing a label above it:

```
$$f(x)\buildrel\hbox{\sevenrm def}\over= {1\over 1+x^2}$$
```

gives

$$f(x) \stackrel{\text{def}}{=} \frac{1}{1+x^2}$$

Notice the curious syntax: `\buildrel...\over{...}`. If the `\over` is followed by a single character or control sequence, you don't need braces, but all hell will break loose if you write something like `\buildrel... \over\hbox{...}`.

Sometimes it's useful to write underneath a symbol, instead of, or in addition to, above it:

$$X_n \xrightarrow[n\to\infty]{\text{weak}} 0 \qquad\qquad \mathcal{P} \xRightarrow[\text{dd}]{*} \mathcal{Q}$$

Let's borrow the definition of `\buildrel` (page 361 of *The TEXbook*) and adapt it to do that. It goes like this:

```
\def\buildrel#1\over#2{\mathrel{
     \mathop{\kern 0pt#2}\limits^{#1}}}
```

We recognize the use of `\mathop`, with the trick of adding something invisible so a single character won't be shifted vertically: see the end of 11.10. We also know that `\limits` makes the superscripted text go above the operator, no matter what the style. The `\mathrel` command transforms the whole thing into a relation, so it'll be spaced right. It's easy enough to rewrite the definition so it takes a subscript as well as a superscript, at the same time making the syntax cleaner:

```
\def\bbuildrel#1_#2^#3{\mathrel{
     \mathop{\kern 0pt#1}\limits_{#2}^{#3}}}
```

Now the two expressions above can be easily coded:

```
X_n\bbuildrel\hbox to .4in{\rightarrowfill}
  _{n\rightarrow\infty}^{\hbox{\sevenrm weak}} 0

{\cal P}\bbuildrel\Longrightarrow_{\hbox{\sevenrm dd}}
    ^*{\cal Q}
```

When using `\bbuildrel` you must have both the `_` and `^` present, and in this order, since they serve to delimit the arguments. If you don't want a label underneath, you must say `\bbuildrel..._{}^...` (or simply use `\buildrel`); if you don't want a label above, say `\bbuildrel..._...^{}`.

11.15 Combining relations

Suppose you need a symbol for a new relation, and decide to make it by combining a plus, a circle and an arrow: $X \mathbin{+\!\circ\!\to} Y$. Here's what you can define a control sequence, say `\toto`, for your brainchild:

```
\def\relplus{\mathrel+}  \def\relcirc{\mathrel\circ}
\def\totosymb{\relplus\joinrel\relcirc\joinrel\rightarrow}
\def\toto{\mathrel{\totosymb}}
```

You start by transforming the symbols into relations (`\rightarrow` is already one): this way TEX puts no space between them when writing them one after the other. Here's how they would come out otherwise: $+\circ\ \to$. Then you use \joinrel to bring them together a bit: this is for good measure, since the characters may not extend as far right and left as their box boundaries. Finally, you again make the whole into a relation.

As long as we're at it, let's build a longer version $+\!-\!\circ\!\longrightarrow$ for display math (see \mathchoice in the Dictionary to find out how to make T_EX choose between the two versions automatically):

```
\def\relminus{\mathrel-}
\def\Totosymb{\relplus\joinrel\relminus
    \joinrel\relcirc\joinrel\longrightarrow}
\def\Toto{\mathrel{\Totosymb}}
```

To place stuff above and below the new symbol, you can use the macros from the previous section:

$X \toto_{n\to\infty} Y$ .... $X +\!\circ\!\longrightarrow_{n\to\infty} Y$

$$X \bbuildrel\Toto_{n\to\infty}^{} Y$$ .... $X \underset{n\to\infty}{+\!-\!\circ\!\longrightarrow} Y$

11.16 More custom-made symbols: limits

The mathematical operation of passing to the limit is represented by the abbreviation lim:

$\lim_{x\rightarrow 0^+}x\ln x=0$ .... $\lim_{x\to 0+} x\ln x = 0$

$$\lim_{x\rightarrow 0^+}x\ln x=0$$ .... $\lim\limits_{x\to 0+} x\ln x = 0$

Plain T_EX also offers the control sequences \limsup and \liminf, which print as lim sup and lim inf. Some people prefer instead the alternate abbreviations $\overline{\lim}$ and $\underline{\lim}$, and it's easy enough to humor them:

```
\def\limsup{\mathop{\overline{\rm lim}}}
\def\liminf{\mathop{\underline{\rm lim}}}
```

The \mathop is essential to make the new operator behave like the old \lim:

$$\liminf_{x\rightarrow 0}f(x)$$ .... $\underline{\lim}\limits_{x\to 0} f(x)$

There are also inductive and projective limits, represented by the abbreviations lim ind and lim proj or $\underrightarrow{\lim}$ and $\underleftarrow{\lim}$, according to taste. None of these is predefined in plain T_EX. The first pair is easy enough:

```
\def\limind{\mathop{\rm lim\,ind}}
\def\limproj{\mathop{\rm lim\,proj}}
```

The second pair is somewhat trickier, because there are no \underrightarrow and \underleftarrow macros to place extensible arrows under an expression. But we can use \longrightarrow and \longleftarrow, which are more or less the right length—a bit too long, so we'll pad the lim with \hfil s. In order to place the arrow as close as possible to the lim, we borrow a macro that plain T_EX uses for this sort of thing. It's called `\oalign`, and it's basically `\halign` with a trivial one-column template `#\cr` and interline spacing set to a minute amount.

```
\def\limind{\mathop{\oalign{\hfil$\rm lim$\hfil\cr
    $\longrightarrow$\cr}}}
\def\limproj{\mathop{\oalign{\hfil$\rm lim$\hfil\cr
    $\longleftarrow$\cr}}}
```

`$U=\limind_{\,\alpha\in A}U_\alpha$` . . . . $U = \varinjlim_{\alpha \in A} U_\alpha$

`$$U=\limind_{\,\alpha\in A}U_\alpha$$` . . . . $U = \varinjlim_{\alpha \in A} U_\alpha$

`$T=\limproj_{\,\omega\in Z}T_\omega$` . . . . $T = \varprojlim_{\omega \in Z} T_\omega$

`$$T=\limproj_{\,\omega\in Z}T_\omega$$` . . . . $T = \varprojlim_{\omega \in Z} T_\omega$

11.17 Phantoms

What is a phantom? It's something invisible that has the same dimensions as a given formula. Phantoms are like struts, but more flexible.

Recall that struts are invisible rules of width zero but non-zero height and/or depth. We've already met several uses for them, mostly in ensuring consistent spacing and positioning in alignments, under radicals and fraction bars, and so on. There are two predefined struts in plain TEX: `\strut` is 8.5 pt tall and 3.5 pt deep, so it supports a whole line; while `\mathstrut` is exactly as tall and as deep as a set of parentheses. Here they are made visible by lending their dimensions to a `\vrule`:

`( \hbox{\mathstrut\vrule} ) \hbox{\strut\vrule}` . . . . (│) │

The `\vphantom` macro makes a strut with same height and depth as the macro's argument: for instance, the definition of `\mathstrut` is `\vphantom{(}`. The argument can be anything, not just a single character. It doesn't even have to be a math formula: `\vphantom` can be used in any mode.

Similarly, `\hphantom` makes an invisible horizontal rule, without height or depth, and having same width as its argument. In section 9.8 we used `\hphantom{0}` to stand in for a digit. And there's also `\phantom`, which makes a whole box, all of whose dimensions match the dimensions of the argument. Both of them can be used outside math mode.

Remark: all these macros are in a way complementary to `\smash`, which prints its argument but takes away its height and depth. In contrast, `\phantom` keeps the dimensions of its argument but makes it invisible.

As an exercise, typeset the following motto ("Let no one ignorant of geometry enter here") for Plato's Academy. Hint: The whole ensemble was set in a `\vbox` of height 14 pt, with interline spacing turned off. The rules in the middle were obtained with `\line{\hrulefill\hphantom{\motto}\hrulefill}`, after a suitable definition for `\motto`.

$$\equiv \text{ΑΓΕΩΜΕΤΡΗΤΟΣ ΟΥΔΕΙΣ ΕΙΣΙΤΩ} \equiv$$

11.18 Displaying several formulas

We've seen that to get a centered formula you should surrounded it with double dollar signs `$$`. But if you have several formulas in a row, and wrap each one in `$$`, the spacing between them becomes excessive. It is better to stack up all the formulas within one set of `$$`. You can do that with plain TEX's `\displaylines` command, with essentially makes a centered alignment with a single column (so no ampersands are needed).

$$\Gamma(z) = \sum_{n=0}^{\infty} \frac{(-1)^n}{n!\,(n+z+1)} + \int_1^{\infty} e^{-t}t^{z-1}\,dt$$

$$\Gamma(z) = \int_0^{\infty} e^{-t}t^{z-1}\,dt$$

$$\Gamma(z+1) = z\Gamma(z)$$

```
$$\openup2pt \displaylines{
   \Gamma(z) ... dt  \cr
   \Gamma(z) ... dt  \cr
   \Gamma(z+1)=z\Gamma(z)  \cr
}$$
```

Here are a few things to watch out for:

• As in any alignment, each line to be centered should end with `\cr`. Actually, `\displaylines` will supply the `\cr` at the end of the last line if it is missing, but there isn't much point in making use of this feature.

• Any punctuation should come before each `\cr`. punctuation in displaysA period right after a `\cr` will be set at the beginning of the next line. A period after the last `\cr` has even more amusing consequences: TEX naturally assumes that it should be put on a line by itself, so it creates a rather piddling extra "formula." No error message is generated, because a `\cr` is quietly supplied for the spurious last line.

• Another common mistake is to forget the braces that close `\displaylines`. This causes a `Runaway argument?` error message and fouls things up to such a degree that you may need to type several carriage returns before TEX finds its bearings again. As usual, the best way to avoid these headaches is to do things from the outside in: start with a skeleton

```
$$\displaylines{
}$$
```

and only then fill the interior of the `\displaylines`. By getting the formatting out of the way first and then concentrating on the formula itself, you'll be much less likely to make errors.

• If the formulas contain large operators or fractions, you may need to increase the spacing between lines for better visual effect. That's why the example above started with `\openup 2pt`. The `\openup` must come *before* the `\displaylines`;

it'll have no effect inside. See section 11.22 for more details. You can also use \noalign to insert spacing or text between the equations in a \displaylines.

Long formulas

You can also use \displaylines for a displayed formula that doesn't fit on one line, because TeX won't break it for you. You must find a good breakpoint yourself, and work as if there were two formulas. (Displayed formulas are generally broken just before an operator, like $+$ or $-$. If at all possible, choose an operator at the "top level," that is, not inside parenthesis or other subformulas.)

The idea is that \displaylines centers lines using weak springs \hfil, which you neutralize with strong springs \hfill:

$$U_n = a_0b_n + a_1b_{n-1} + \cdots + a_{n-1}b_1 + a_0b_n$$
$$+ c_0d_n + \cdots + c_0d_n + \int_0^\infty \frac{R_n(t)}{1+t^2}dt$$

```
$$\displaylines{
   \qquad  U_n=a_0b_n+a_1b_{n-1}+\cdots
      +a_{n-1}b_1+a_0b_n\hfill\cr
   \hfill {}+c_0d_n+\cdots+c_0d_n+
      \smash{\int_0^\infty{R_n(t)\over 1+t^2}dt}\qquad\cr
}$$
```

There are several typographical niceties in this display, all of them worth preserving in similar situations:

• The \qquad at the beginning of the first half and at the end of the second prevents the two halves of the formula from touching the margins. If the formula is really long you can reduce it to one \quad, or, as a last resort, do without it altogether.

• The large integral sign in the second half is not directly below the first half, but it makes the line it's on taller than usual anyway. If we had taken no precautions, there would seem to be extra spacing between the lines. Here a well-placed \smash make TeX ignore the extra height. (Section 8.8 showed another example of the same problem.)

• The code for the second half starts with {}+c_0d_n+... (not counting the \hfill). What is the empty group {} doing there? It turns out that when TeX sees a binary operator at the beginning of a formula, it assumes that it's a unary operator instead, and doesn't put any spacing around it:

$+x$, ${}+x$ .... $+x, +x$

This is because $+$ and $-$ can serve both functions, and the only way TeX has to know the difference is to see if there are expressions on both sides. Here, in spite of appearances, $+$ is binary, and we made that clear by starting the line with the dummy. You should do likewise whenever you split a formula at any binary operator, such as $\times$ or $*$.

11.19 Aligning several formulas

Rather than centering each formula in a multi-line display, it is often preferable to align them by their $=$ signs or some other convenient character. This is done with the \eqalign macro:

$$
\Gamma(z) = \sum_{n=0}^{\infty} \frac{(-1)^n}{n!\,(n+z+1)} + \int_1^{\infty} e^{-t} t^{z-1}\, dt
$$

$$
\Gamma(z) = \int_0^{\infty} e^{-t} t^{z-1}\, dt
$$

$$
\Gamma(z+1) = z\Gamma(z)
$$

```
$$\openup2pt\eqalign{
    \Gamma(z)& =\sum ... dt \cr
    \Gamma(z)& =\int ... dt \cr
    \Gamma(z+1)& =z\Gamma(z) \cr
}$$
```

An ampersand & separates the left and right parts of each line, and a \cr terminates each line. In other words, the contents of an \eqalign{...} are like the rows of an \halign with two columns. The $=$ signs, or any other characters, following the ampersands are aligned vertically.

Most of the comments and warnings we made about \displaylines apply here too, but there is one difference: \eqalign is unresponsive to springs. Writing A\hfill&B\cr or A&\hfill B\cr as a row of an \eqalign is the same as writing A&B\cr. The reason is given in section 12.10, together with alternatives.

The left side of one or more formulas of an \eqalign can be empty:

$$
(x+1)^3 - (x-1)^3 = x^3 + 3x^2 + 3x + 1 - (x^3 - 3x^2 + 3x - 1)
$$
$$
= 2(x^2 + 1)
$$

```
$$\eqalign{
    (x+1)^3-(x-1)^3&=x^3+3x^2+3x+1-(x^3-3x^2+3x-1)\cr
        &=2(x^2+1)\cr}
$$
```

If all the rows have nothing before the & , the formulas will be left-aligned. Right-aligning formulas is equally easy. This might seem obvious, but it's surprising how many beginners find themselves at a loss to align formulas on one side!

There are other unexpected ways to use \eqalign. The arrangement on the next page shows how naturally TₑX's math facilities can be harnessed for apparently unrelated uses. To obtain it, we first put the two quotations and author names in boxes:

```
\setbox1=\vbox{\hsize 3.5in Farewell, eyes that I loved!... }
\setbox2=\vbox{\hbox{Antoine de}\hbox{Saint-Exup\'ery}}
\setbox3=\vbox{\hsize 3.5in  Now this, monks... }
\setbox4=\hbox{Buddha}
```

Then it was easy enough to put everything together in math mode:

```
$$\openup 6pt\eqalign{
    &\left.\vcenter{\box1}\right\}\vcenter{\box2}\cr
    &\left.\vcenter{\box3}\right\}\vcenter{\box4}\cr
}$$
```

Farewell, eyes that I loved! Do not blame me if the human body cannot go three days without water. I would never have thought myself so truly a prisoner of springs. I had no notion that my self-sufficiency was so circumscribed. We take it for granted that a man is able to stride straight out into the world. We believe that man is free. We never see the cord that binds him to the well, that umbilical cord by which he is tied to the womb of the world. Let man take but one step too many... and the cord snaps.[1]

Antoine de Saint-Exupéry

Now this, monks, is the noble truth of the cause of pain: the thirst that tends to rebirth, combined with pleasure and lust, finding pleasure here and there; the thrist for passion, the thrist for existence, the thirst for non-existence.[1]

Buddha

You can have other material together with \eqalign in a display. To separate lines, you can use \openup, as shown above, or \noalign. But \noalign won't help in inserting text between the formulas in an \eqalign; see the Dictionary under \eqalignno for a better idea.

11.20 Labeling formulas

To label a single formula, like this,

$$\frac{e^{ux} - e^{u}}{e^{u} - 1} = x - 1 + \sum_{1}^{\infty} B_n(x)\frac{u^n}{n!} \tag{1}$$

place \eqno and the label after it:

```
$$
{e^{ux} ... {u^n \over n!} \eqno (1)
$$
```

Everything after the \eqno turns into the label. To place a label against the left margin, use \leqno instead of \eqno; the position of \leqno stays the same:

```
$$
{e^{ux} ... {u^n \over n!} \leqno (1)
$$
```

---

[1] From *Terre des Hommes*, based on the translation of Lewis Galantière titled *Wind, Sands and Stars*.

[2] From the Sermon at Benares, based on E. A. Burtt in *The Teachings of the Compassionate Buddha*.

To label several formulas you might try to combine `\eqalign` with `\eqno`. This works fine if you want the formulas labeled as a group: the label is placed halfway down the alignment, and, if necessary, you can indicate explicitly that it refers to all of the formulas by using braces:

$$\left.\eqalign{\Gamma(z) &= \sum_{n=0}^{\infty} \frac{(-1)^n}{n!\,(n+z+1)} + \int_1^{\infty} e^{-t}t^{z-1}dt \cr \Gamma(z) &= \int_0^{\infty} e^{-t}t^{z-1}\,dt \cr \Gamma(z+1) &= z\Gamma(z) \cr}\right\} \qquad (17)$$

```
$$\openup 2pt\left.\eqalign{
   \Gamma(z)& =\sum ... dt \cr
   \Gamma(z)& =\int ... dt \cr
   \Gamma(z+1)& =z\Gamma(z) \cr
}\right\}
\eqno (17)$$
```

But if you need to label one or more equations individually, this won't do. You must instead use the variants `\eqalignno` and `\leqalignno`. Here are the Newton–Girard formulas, individually numbered:

$$S_1 + a_1 = 0 \qquad (1)$$
$$S_2 + S_1 a_1 + 2a_2 = 0 \qquad (2)$$
$$\cdots\cdots\cdots\cdots\cdots\cdots\cdots$$
$$S_n + S_{n-1}a_1 + S_{n-2}a_2 + \cdots + S_1 a_{n-1} + na_n = 0 \qquad (n)$$

```
$$\eqalignno{
   S_1+a_1=0&& (1) \cr
   S_2+S_1a_1+2a_2=0&& (2) \cr
   \hbox to 2in{\dotfill}& \cr
   S_n+S_{n-1}a_1+S_{n-2}a_2+\cdots+S_1a_{n-1}+na_n=0&& (n) \cr
}$$
```

As you see, each row now has two ampersands: one to determine where the formulas align, and one to delimit the labels. If there is no label, you don't need the corresponding `&`.

By moving the first ampersand on each line to the beginning and by replacing `\eqalignno` with `\leqaligno`, we get the mirror image arrangement:

$$(1) \qquad S_1 + a_1 = 0$$
$$(2) \qquad S_2 + S_1 a_1 + 2a_2 = 0$$
$$\cdots\cdots\cdots\cdots\cdots\cdots\cdots$$
$$(n) \qquad S_n + S_{n-1}a_1 + S_{n-2}a_2 + \cdots + S_1 a_{n-1} + na_n = 0$$

Plain TeX doesn't offer a macro to number the formulas of a `\displaylines`. If you try doing it with `\eqno`, it chokes. In section 12.10 we'll plug this gap with the `\displaylinesno` and `\ldisplaylinesno` macros.

11.21 Matrices

Not surprisingly, `\matrix` is the macro that makes matrices, those arrays of numbers, letters and formulas that mathematicians are so fond of:

```
$$\matrix{
  \alpha           &\beta&\gamma\cr
  \cal A           &B    &C     \cr
  x_1+\cdots+x_\ell &y    &z     \cr
}$$
```

$$\matrix{
\alpha & \beta & \gamma\cr
\mathcal{A} & B & C\cr
x_1 + \cdots + x_\ell & y & z\cr
}$$

This syntax is by now familiar: ampersands `&` separate the entries in each row, while `\cr` terminates each row, including the last. If the rows have different numbers of ampersands, the matrix will have as many columns as the longest row, and the shorter ones will be filled with empty entries, as if they ended with `...&&&\cr` (cf. paragraph 9.4).

Matrices are a particular case of alignments, so much of what we said in chapter 9 is relevant here. First, each entry forms a group; for example, the `\cal` command in row 2, column 1 of the matrix above has no effect on other entries. Second, you can't have a group straddling several entries, since groups must nest properly. Third, you can add rules and spacing between rows by using `\noalign`. (But `\openup` and `\offinterlineskip` won't work, because `\matrix` returns the interline spacing to its default value; see section 11.22.) Finally, the entry templates are basically `\hfil$#$\hfil`, so each entry is read in math mode and set in text style, centered in its column.

As usual, mistakes will be fewer likely if you type your matrices from the outside in, starting with the braces and only then filling in the entries. It is also good to keep columns aligned in the source file, if possible; since entries are read in math mode, there is no danger of spurious spaces showing on the output.

Matrices and springs

Since the centering of entries is achieved with weak springs, you can use `\hfill` to left-align or right-align columns:

```
$$\matrix{
  \hfill x  &y\hfill  &z     \cr
  \hfill x' &y'\hfill &z'    \cr
  \hfill x''&y''\hfill&z''   \cr
}$$
```

$$\matrix{
x & y & z\cr
x' & y' & z'\cr
x'' & y'' & z''\cr
}$$

Matrices in parentheses

While `\matrix` creates a naked matrix, `\pmatrix` dresses its results in parentheses. In this example, two of the entries of the big matrix are themselves matrices:

$$J = \begin{pmatrix} \begin{pmatrix} \lambda & 1 \\ 0 & \lambda \end{pmatrix} & \mathbf{0} \\ \mathbf{0} & \begin{pmatrix} \mu & 1 & 0 \\ 0 & \mu & 1 \\ 0 & 0 & \mu \end{pmatrix} \end{pmatrix}$$

```
$$J=\pmatrix{
\pmatrix{\lambda&1\cr
         0       &\lambda\cr}     &\bf 0\cr
\bf 0&\pmatrix{\mu&1     &0  \cr
               0 &\mu &1< \cr
               0 &0    &\mu\cr}         \cr
}$$
```

TEX automatically centers the small matrices vertically, with respect to the **0** on the same row. It also centers the big matrix with respect to the $J =$.

Determinants

The determinant of a matrix is represented by clothing the matrix with vertical bars, instead of parentheses. This is easy to do with the \left|...\right| construction of section 11.13:

$$\det(A - \lambda I) = \begin{vmatrix} a_{11} - \lambda & a_{12} & a_{13} \\ a_{21} & a_{22} - \lambda & a_{23} \\ a_{31} & a_{32} & a_{33} - \lambda \end{vmatrix}$$

```
$$\det(A-\lambda I)=\left|\matrix{
  a_{11}-\lambda&a_{12}\hfill&a_{13}\hfill\cr
  a_{21}\hfill&a_{22}-\lambda&a_{23}\hfill\cr
  a_{31}\hfill&a_{32}\hfill&a_{33}-\lambda\cr
}\right|$$
```

Here it seemed better to align entries by their common letter, rather than center them, so we used \hfill to left-justify.

Systems of equations

Systems of equations provide another possible use for \matrix:

$$(\Sigma) \qquad \left\{ \begin{matrix} 2x & + & 3y & - & 45z & = & b_1 \\ -12x & - & 41y & + & z & = & b_2 \\ & & 6y & + & 9z & = & b_3 \end{matrix} \right.$$

```
$$\left\{\matrix{
  \hfill 2x &+&\hfill 3y  &-&        45z &=& b_1\cr
      -12x &-&        41y &+& \hfill  z &=& b_2\cr
        & & \hfill 6y &+& \hfill 9z &=& b_3\cr
}\right. \leqno (\Sigma)$$
```

Since there is no delimiter to match the left brace, we use the dummy `\right`. (with a period after the `\right`): see section 11.13. Here we choose to right-justify the entries, so the variables x, y and z are aligned. The label of an equation, as you see, doesn't have to be a number. TₑX reads the material following `\eqno` and `\leqno` in math mode, so we don't need `$...$` around the `\Sigma`.

We will write in section 12.10 a `\system` macro that automates the coding of systems of equations somewhat, and does a better job with the spacing.

11.22 Adjusting the spacing

Sometimes it is desirable to change the spacing between lines in a display, mostly often to open it up. The situation varies depending on the macro used to form the lines, and on whether all lines should be separated or just two.

In sections 11.18 and 11.19 we saw how `\openup` can be used to separate all the lines of a `\displaylines` or `\eqalign`. The same works with `\eqalignno` and `\leqalignno`. The important thing to remember is that `\openup` must be used before the macro that creates the alignment, not inside it:

$$\openup\ 4pt\displaylines\{...\}$$

The effects of `\openup` accumulate: `\openup 2mm` followed by `\openup 3mm` gives the same result as `\openup 5mm`. You can also use a negative dimension to bring lines closer together.

With `\matrix`, `\pmatrix`, `\cases` and some other macros, `\openup` won't work: the first thing these macros do is reset the interline spacing to its default values. Plain TₑX stores those values in the variables `\normalbaselineskip`, `\normallineskip` and `\normallineskiplimit`. To control the interline spacing in a matrix, those are the variables we must change, rather than changing `\baselineskip`, `\lineskip` and `\lineskiplimit` directly.

It's best to make any changes to the default values inside a group—otherwise, they'll affect the whole document. In the next few sections, then, when we want TₑX to try to keep baselines separated by 15 pt, we'll say

$$\{\normalbaselineskip=15pt\matrix\{...\}\}$$

Separating two lines

All these maneuvers are designed to open up an alignment by separating all their lines. But in practice one also wants to separate two consecutive lines, leaving the others alone. The solution in this case is the same for all types of alignment: inserting `\noalign{\vskip...}` between two rows, that is, right after a `\cr`:

```
$$\eqalign{
... & ... \cr
\noalign{\medskip}
... & ... \cr
}$$
```

This will separate lines by an extra 6 pt. To pull lines together, use a negative dimension: `\noalign{\vskip -3pt}`.

11.23 Ellipses

If you type three dots to indicate an ellipsis, TEX prints them too close: ... As a discriminating user you will prefer to use the `\dots` command, which gives ... .

In mathematics there are other arrangements for the three dots, and consequently other control sequences. Here's the list:

- `\ldots` gives low dots, like `\dots` (the difference isn't worth fretting about). This version is used between commas and other punctuation:

`$x=(x_1,\ldots,x_n)$` .... $x = (x_1, \ldots, x_n)$

- `\cdots` gives dots at the level of a $+$, so this version looks best between operators like $+$ and $\times$:

`$S=x_1+\cdots+x_n$` .... $S = x_1 + \cdots + x_n$

- `\vdots` and `\ddots` give dots arranged vertically and diagonally, respectively; they're mostly used with matrices and other alignments:

```
$$H=\pmatrix{
   a_{11}&a_{12}&\cdots&a_{1n}\cr
   a_{21}&a_{22}&\cdots&a_{2n}\cr
   \vdots&\vdots&\ddots&\vdots\cr
   a_{n1}&a_{n2}&\cdots&a_{nn}\cr
}$$
```

$$H = \begin{pmatrix} a_{11} & a_{12} & \cdots & a_{1n} \\ a_{21} & a_{22} & \cdots & a_{2n} \\ \vdots & \vdots & \ddots & \vdots \\ a_{n1} & a_{n2} & \cdots & a_{nn} \end{pmatrix}$$

On top of all this, sometimes you need diagonal dots $\cdot^{\cdot^{\cdot}}$ running in the opposite direction, which plain TEX doesn't offer. No problem; we just crib the definition of `\ddots` from page 359 of *The TEXbook*, and define our very own `\adots` macro by switching around the endpoints (the 'a' is for ascending):

```
\def\adots{\mathinner{\mkern2mu\raise1pt\hbox{.}\mkern2mu
   \raise4pt\hbox{.}\mkern2mu\raise7pt\hbox{.}\mkern1mu}}
```

`$\adots\ddots\ldots\adots\ddots$` ....

Here's a more serious application:

$$M = \begin{pmatrix} a_{11} & a_{12} & \cdots & a_{1,n-1} & 1 \\ a_{21} & a_{22} & \cdots & 1 & 0 \\ \vdots & \vdots & \cdot^{\cdot^{\cdot}} & 0 & 0 \\ a_{n-1,1} & 1 & 0 & 0 & 0 \\ 1 & 0 & 0 & 0 & 0 \end{pmatrix}$$

The only noteworthy things here are the `\adots` on the third row, and the use of `\normalbaselineskip` to open up the matrix, as explained in the previous section (no extra level of grouping is needed here, because the $$ already group):

```
$$\normalbaselineskip=15pt
M=\pmatrix{ ...
   \vdots & \vdots & \adots & 0 & 0 \cr
   ... }$$
```

11.24 Diagrams

A diagram makes a mathematician's day, especially if it is commutative! But it takes some experience to get TeX to arrange all those horizontal and vertical arrows in their right places, as well as formulas above, below and next to the arrows. If you feel ready, take a long breath, and here it goes:

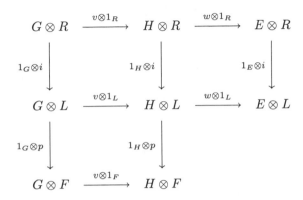

To make this diagram, we started with an abbreviation, `\def\ot{\otimes}`. We also used three macros to simplify the coding, whose definitions are given later in this section, as they're not part of plain TeX:

- `\diagram` is a variant of `\matrix`, and is used exactly the same way.

- `\harr` makes an arrow pointing right, half an inch in length. It takes two arguments in braces, and places the first above the arrow and the second below, both in script style. In this case we had use for only one label per arrow, but we had to use two sets of braces anyway: `harr{v\ot1_F}{}`. Otherwise TeX would commandeer the next thing in line to be an argument, and chaos would be guaranteed to ensue.

- `\varr` is very similar: it draws an arrow pointing down, also half an inch long. Again, we had to supply two arguments in braces, one or them empty in this case. The first argument is placed to the left, and the second to the right of the arrow, both in script style.

With these macros, coding the diagram is straightforward, if not exactly a fascinating task:

```
$$\diagram{
  G\ot R&\harr{v\ot1_R}{}&H\ot R&\harr{w\ot1_R}{}&E\ot R\cr
  \varr{1_G\ot i}{}&&\varr{1_H\ot i}{}&&\varr{1_E\ot i}{}\cr
  G\ot L&\harr{v\ot1_L}{}&H\ot L&\harr{w\ot1_L}{}&E\ot L\cr
  \varr{1_G\ot p}{}&&\varr{1_H\ot p}{}&&\cr
  G\ot F&\harr{v\ot1_F}{}&H\ot F\cr
}$$
```

Here's another diagram which, although simpler, has some interesting subtleties:

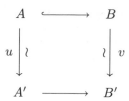

The arrow at the top is obtained with `\lhook\joinrel\mathrel{\harr{}{}}`. Notice also the use of `\displaystyle` to set the labels in 10 point, and the use of `\kern-2pt` to get the `\wr` closer to the arrows:

```
$$\diagram{
  A & \lhook\joinrel\mathrel{\harr{}{}} & B \cr
  \varr{\displaystyle u}{\kern-2pt\displaystyle\wr} &&
  \varr{\displaystyle\wr\kern-2pt}{\displaystylev} \cr
  A'& \harr{}{} & B' \cr
}$$
```

Let's turn now to the three macros used above: `\harr`, `\varr` and `\diagram`. The horizontal arrow is easy to define:

```
\def\harr#1#2{\smash{\mathop{\hbox to .5in{\rightarrowfill}}
  \limits^{\scriptstyle#1}_{\scriptstyle#2}}}
```

Notice the use of `\smash` so the height and depth of the labels will not interfere with the spacing between rows in the diagram. The vertical arrow is almost as easy:

```
\def\varr#1#2{\llap{$\scriptstyle #1$}\left\downarrow
  \vcenter to .5in{}\right.\rlap{$\scriptstyle #2$}}
```

At end of 11.13 we saw why `\left\downarrow\vcenter to .5in{}\right.` produces a vertical arrow exactly half an inch long. The labels for the arrow is then written with `\llap` and `\rlap` so TₑX won't see their width; this way they won't disturb the alignment.

We could almost do without `\diagram`, since `\matrix` is close to being perfect for the task: it reads each entry (including the vertical arrows) in math mode, and centers it in its column. It has one defect, however—it starts by restoring plain TₑX's default interline spacing. This is insufficient to separate the vertical arrows from the letters: the arrows are tall and deep, and according to the rules of section 8.9, this means that TₑX will use the `\lineskip` of 1 pt to separate rows.

Our goal, then, is to set the `\lineskip` to a more reasonable value, like 8 pt, and also to make the `\baselineskip` zero so as to ensure that the `\lineskip` will always be used. As we discussed in section 11.22, we must do that indirectly, assigning the desired value to `\normallineskip`, rather than to `\lineskip`;

and we must do it in inside a group, so the change is reversible. With that in mind, the definition of \diagram is a cinch:

```
\def\diagram#1{{\normallineskip=8pt
    \normalbaselineskip=0pt \matrix{#1}}}
```

You should have no trouble adapting these definitions to force a different amount of interline spacing, or to generate arrows of different sizes, or pointing the other way. Unfortunately, plain TeX can't draw diagonal arrows of variable length, so you can't use them in diagrams.[3]

---

[3] LATeX has a wider, but still limited, selection of arrows. Other macro packages, like PicTeX, define more general graphics commands, but it's still far from trivial to draw a complicated diagram using them.

12
T_EX Programming

In the first eleven chapters of this book we encountered numerous macros, and even created some, always more or less informally. It is now time to be a bit more systematic, and cover the basics of TEX programming. Our study will by no means be exhaustive—a whole separate book could be written on the subject—but we hope by the end of the chapter you'll have a solid understanding of what a macro is and of what happens when TEX stores a macro and when it uses it.

12.1 Generalities

Control sequences

When TEX reads a backslash \ , it knows that following word is not to be printed, but treated as a command. A word starting with a backslash is called a control sequence. What follows the backslash can be either an arbitrary number of letters (lowercase or uppercase), or a single non-letter. A control sequence made of letters ends just before the first non-letter.

There are four control sequences in the string \%\toto(5)\tata xxx\titi21 . The first is \% ; it's an example of a control sequence containing a single non-letter. The second is \toto , which ends just before the left parenthesis. The third is \tata , which ends with a space; this is the commonest way to mark the end of a control sequence. Finally, there is \titi , whose end is announced by a digit.

It follows that the name of a control sequence can't contain both letters and digits. Sometimes one wishes this weren't so, but one can always use roman numerals: \LouisXVI . Notice also that TEX distinguishes between upper- and lowercase letters, so \toto and \Toto are distinct control sequences.

After reading a control sequence made of letters TEX will ignore all spaces and up to one carriage return until it finds something else.

Defining a macro

Control sequences are basically of two types: primitives and macros. Primitives are the building blocks of TEX, direct commands for TEX's engine. They have a meaning when TEX starts up, even before it's read the `plain.tex` file that defines the plain TEX format. A macro, on the other hand, has no predefined meaning. Its meaning is assigned by a definition, which is introduced by the `\def` primitive:

$$\def\toto{\quad{\it TOTO\/}}$$

When TEX sees `\def\toto`, it stores `\toto` in a dictionary, together with its *replacement text*, which is everything that follows in braces. TEX doesn't try to execute or understand the replacement text while it's memorizing it; in fact it barely notices what it contains, except that it keeps an eye out for braces, so it knows where to stop. A replacement text must have balanced braces, so in the example above TEX doesn't stop till it reaches the second `}`; if it stopped at the first, the replacement text would be `\quad{\it TOTO\/`, which is unbalanced.

Expanding a macro

When TEX encounters a control sequence that's in the macro dictionary, it proceeds to *expand* it. This means that it replaces the control sequence by its replacement text, and starts reading the text obtained in this way. In our example, `\toto` would be replaced by `\quad{\it TOTO\/}`, and TEX would start reading again from `\quad`. Now `\quad` is itself a macro, defined in `plain.tex`, so TEX would replace `\quad` with its definition:

$$\hskip 1em\relax{\it TOTO\/}$$

No further expansion takes place now, because `\hskip` is a primitive; TEX reads its complement `1em`, and executes the command. It continues with `\relax`, also a primitive (see the Dictionary), and with the `{`, which makes it start a new group. Next it sees `\it`, which is again a macro; it replaces it by its definition, and so on.

There are other, less common, types of control sequences: font names (section 4.3), clones (section 12.2), register names (section 12.8), and others. For our current purposes we can think of them as primitives, because they are executed, rather than expanded: a font name, for instance, is a command to change the current font. (But a clone of a macro will be expanded.)

You can define a new macro using `\def` anywhere. If you do it inside a group, the definition will disappear at the end of the group, unless you precede `\def` with `\global`. The primitive `\gdef` is an abbreviation for `\global\def`.

A macro can be defined many times over; each definition erases the previous one. Since TEX won't warn you that you're about to clobber an existing definiton, you can get into trouble if you're not careful. But later on we'll see a way to protect a macro so it cannot be clobbered.

Don't abuse the right to define macros anywhere. If you do it all over the place, your file will become impenetrable. It's much better to group all your definitions (except perhaps those that you only need very briefly, inside a group) in a separate file, as explained in section 1.7.

12.2 Abbreviations and clones

Macros like \toto and \quad in the previous paragraph are essentially abbreviations: they always expand to the same thing. Abbreviations are easy to define, and you'll probably find yourself using them quite a bit. If "two-dimensional" and "WYSIWYG" appear several times per page in your document, you'll save time and avoid errors by setting

```
\def\twod{two-dimen\-sional} \def\wysiwyg{{\eightrm WYSIWYG}}
```

and typing \twod and \wysiwyg after that.

We chose these definitions because they make important points. The \- in the first is called a discretionary hyphen, and it tells TEX that it can hyphenate the word at that point. If you don't specify discretionary hyphens, TEX sticks to the old-fashioned rule, which we just broke, that compound words should not be further hyphenated.

In the second definition, notice the two sets of braces. The outer one merely delimits the replacement text; it does not become part of it. Without the inner braces, TEX will see \eightrm WYSIWYG when you use the abbreviation—and everything from there on will be set in eight point!

One drawback of using an abbreviation in the middle of text is that you have to write \ after it in order to leave a space: \twod graphics gives two-dimensionalgraphics. You might try instead to put the space inside the replacement text: \def\twod{two-dimen\-sional }. But then it would crop up in unwanted places, like before punctuation.

You can solve the problem at the cost of one extra keystroke. If you say

```
\def\twod/{two-dimen\-sional}
```

TEX will expect to always see / after \twod, and it replaces \twod/ by the replacement text. So you can think of \twod/ as a slightly longer abbreviation. Spaces are treated normally after /, so the problem of deciding when to add an explicit space goes away. And if you forget the /, TEX will give an error message, rather than silently gobbling up the space as before. (Such typos have a way of not being detected till your paper has been distributed widely...)

So much for abbrevations of words. Naturally, you can also abbreviate commands or sequences of commands. Even an abbreviation for a single control sequence, like \def\ot{\otimes} in section 11.24, can be useful if the control sequence name is long or used several times. Think of \Longleftrightarrow!

There is an alternative for one-control-sequence abbreviations, involving TEX's \let primitive. If you say \let\ot=\otimes (the = is optional), you assign to \ot the meaning of \otimes, effectively cloning \otimes. The two control sequences become synonymous.

There is, however, an important difference between the two constructions: with \let\ot\otimes, you're assigning to \ot the *current meaning* of \otimes, so if at a later time the meaning of \otimes changes, that has no effect on \ot.

But with `\def\ot{\otimes}`, you're saying that `\ot` should be replaced by `\otimes` whenever it occurs, so the meaning of `\otimes` at the time of the replacement matters.

You can even clone a character! For example, plain TEX says `\let\bgroup{` and `\let\egroup}`, so `\bgroup` and `\egroup` are just like braces everywhere. Well, almost everywhere: inside a macro definition they're different, because when TEX is storing a definition in its dictionary, it only uses real braces, not cloned braces, to figure out where the definition ends. Thus `\bgroup` and `\egroup` provide a way to have unbalanced "braces" inside a definition. The need for that is easy to see:

Say you want `\narrow` to have the effect of `{\leftskip=1em\rightskip=1em` and `\endnarrow` to have the effect of `\par}`, so paragraphs placed between `\narrow` and `\endnarrow` are indented on both sides. (Compare plain TEX's `\narrower` command.) If you try

        \def\narrow{{\leftskip=1em\rightskip=1em}
        \def\endnarrow{\par}}

TEX matches the *first* left brace with the *last* right brace, and the upshot is that `\narrow` is defined to mean

        {\leftskip=1em\rightskip=1em} \def\endnarrow{\par}

Not at all what you wanted! But everything works out nicely if you say instead

        \def\narrow{\bgroup\leftskip=1em\rightskip=1em}
        \def\endnarrow{\par\egroup}

12.3 Macros with arguments

Even more useful than abbreviations are macros whose replacement text contains variables, or arguments. If a paper contains dozens of formulas like $(x_1, \ldots, x_n)$ or $(y_1, \ldots, y_n)$, we'll start it with the definition

        \def\nuple#1{(#1_1,\ldots,#1_n)}

and say `$\nuple x$` to get $(x_1, \ldots, x_n)$. The idea is simple: every time we say `\nuple`, TEX consults its dictionary and finds that the macro expects an argument. The argument is whatever follows `\nuple`, and it gets plugged into the replacement text in place of every `#1`. So `\nuple x` expands into `(x_1,\ldots,x_n)`, while `\nuple y` expands into `(y_1,\ldots,y_n)`, and so on.

The argument doesn't have to be a single character, but it is unless you say otherwise. To say otherwise, put it in braces: `$\nuple{\overline{AB}}$` gives $(\overline{AB}_1, \ldots, \overline{AB}_n)$. More precisely, if the first character after the macro (not counting spaces and carriage returns) is neither `{` nor `\`, it becomes the argument. If the first character is `{`, the argument is the text between this `{` and the matching `}`. If the character is `\`, the argument is a control sequence.

It's important to realize that *the braces themselves are not part of the argument*, just as the braces that delimit the replacement text in a definition are not part of it.

Forget this, and you'll be in for trouble: `$\nuple{\bf x}+\nuple y$`, for example, will give $(\mathbf{x_1}, \ldots, \mathbf{x_n}) + (\mathbf{y_1}, \ldots, \mathbf{y_n})$: the replacement text after argument substitution is

$$\text{\$(\bf x\_1,\ldots,\bf x\_n)+(y\_1,\ldots,y\_n)\$}$$

so the effects of the first `\bf` last till the end of formula! To get $(\mathbf{x}_1, \ldots, \mathbf{x}_n) + (y_1, \ldots, y_n)$, you must type `$\nuple{{\bf x}}+\nuple y$`.

Now suppose that our paper has not only things like $(x_1, \ldots, x_n)$ and $(y_1, \ldots, y_n)$, but also $(x_1, \ldots, x_p)$ (with a different last index), $(y_1, \ldots, y_q)$ or $(z_1, \ldots, z_{r+s})$. We certainly don't want to have to define a plethora of macros `\puple`, `\quple`, `\rsuple`! Instead we say

$$\text{\def\uple\#1\#2\{(\#1\_1,\ldots,\#1\_\{\#2\})\}}$$

This new macro expects two arguments. Argument 1 is the first character or control sequence or group after the macro, and it gets plugged into the replacement text wherever there is a `#1`. Argument 2 is the first character or control sequence or group after the end of argument 1, and it replaces every occurrence of `#2`. In order to get $(x_1, \ldots, x_p)$, then, it's enough to type `$\uple xp$`: argument 1 is `x` and argument 2 is `p`. To get $(z_1, \ldots, z_{r+s})$, type `$\uple z{r+s}$`: argument 1 is `z` and argument 2 is `r+s`, since what comes after argument 1 is a group. When in doubt, you can use braces: `\uple{x}{p}` and `\uple{z}{r+s}` will work just as well as `\uple xp` and `\uple z{r+s}`. It's better to err on the side of caution.

Notice the braces around `#2` in the replacement text of `\uple`. They're needed in order for `\uple z{r+s}` to work right; if they weren't there the expansion of `\uple z{r+s}` would be $(z\_1,\ldots,z\_r+s)$, which gives $(z_1, \ldots, z_r + s)$. On the other hand, `\uple z{{r+s}}` would work fine. This is exactly the same problem we encountered above with `\nuple{\bf x}`. In fact, we could have made the definition of `\uple`

$$\text{\def\uple\#1\#2\{(\{\#1\}\_1,\ldots,\{\#1\}\_\{\#2\})\}}$$

and then `\uple{\bf x}{r+s}` would work right. Placing braces inside the replacement text is generally a good idea if you don't want to rely on the arguments being watertight; but it isn't always possible, or desirable, to transfer the burden of grouping in this way.

A macro can have up to nine arguments, referred to by `#1`, ..., `#9`. To define a macro with three arguments, you must say `\def\toto#1#2#3{...}`; no flights of fancy like `\toto#2#1#3` or `\toto#1#4#5`. In using a macro, it's crucial to give it all the arguments it expects, and in the right order. If one of the arguments should be replaced by nothing, you must use the empty group `{}` to indicate that fact. We've seen the need for that before: for example, the `\item` macro of section 6.9 expects one argument. If you have nothing to write on the margin, you must say `\item{}...`; otherwise T_EX will take the next character or group for an argument.

Another thing to watch out for is spurious spaces. If the definition of a macro is more than one line long, it's best to end each line with `%`, unless the last thing on it is a control sequence. Otherwise the carriage return—turned into a space—remains in the replacement text, and can affect the output if TEX is in horizontal mode at the time it sees it. The macros in sections 4.7 and 8.11 take this precaution.

Before the replacement text, too, spaces can spell trouble. Of the four definitions

```
\def\toto#1#2{...}   \def\toto #1#2{...}
\def\toto#1#2 {...}  \def\toto#1 #2{...}
```

the top two have exactly the same effect (since a space is ignored after a control sequence), but the second two are very different from the first two and from one another. In them, the space serves to delimit the preceding argument, by a mechanism that we'll discuss in detail in the next section.

12.4 Fine points of macro syntax

in replacement texts

Since a # in macro definition indicates a slot for an argument, what do you do if you want a literal # in the replacement text? You say ##. The most common situation is when the macro is supposed to expand to an alignment command, preamble and all. Many of plain TEX's alignment macros, like \eqalign, \cases, etc., are coded in this way (see section 12.10). Here's a simple example:

```
\def\toto#1{\halign{\bf##&&\quad\hfil##\hfil\cr#1}}
```

If TEX sees \toto{A&B\cr a&b\cr}, it proceeds as usual, plugging A&B\cr a&b\cr in place of #1 in the definition. The replacement text after substitution is

```
\halign{\bf#&&\quad\hfil#\hfil\cr A&B\cr a&b\cr}
```

which is a complete alignment. The users of \toto don't have to know about preambles, ##, or anything of the sort—all of that is tucked away inside the macro definition. They just use \toto as they might use \eqalign:

```
\toto{
  \it function&\it continuous&
     \it periodic\cr
  sine       &yes &$2\pi$\cr
  tangent    &yes &$\pi$ \cr
  polynomial&yes &no    \cr
}
```

function	*continuous*	*periodic*
sine	yes	2π
tangent	yes	π
polynomial	yes	no

Another situation when a ## is necessary is when the replacement text contains another macro definition:

```
\def\TOTO#1#2{...\def\toto##1##2##3{...}...}
```

You must use ## for the arguments of the inner macro even if the outer one doesn't have arguments.

Delimited arguments

Suppose you want to typeset a bibliography with names in caps and small caps (KNUTH), but don't have the appropriate font. Not to worry! You can make do with the ersatz macro

```
\def\csc#1#2!{{\tenrm #1\sevenrm #2}}
```

which you use like this: `\csc KNUTH!` for K<small>NUTH</small>. The definition of `\csc` shows a new twist: a *delimited argument*. The arguments of `\csc` are determined as follows:

- The first, `#1`, is undelimited, since it's immediately followed by `#2` in the definition. An undelimited argument is determined by the rule of the previous section: it's the first character, control sequence or group. In this case it's just the `K`.

- The second argument, `#2`, is delimited by `!`, that is, it is followed in the definition by `!`. This means that it will consist of everything from the end of the first argument up to, but not including, the first `!`. *The delimiter does not become part of the argument*, nor is it part of the replacement text; that's why no exclamation mark appears on the output. In this case the second argument is `NUTH`.

It follows, then, that the replacement text of `\csc KNUTH!` after substitution is `{\tenrm K\sevenrm NUTH}`.

When T_EX sees a macro with delimited arguments it expects to find the delimiters somewhere. Delimiters inside braces don't count, so you can get K<small>NUTH</small>! by saying `\csc KNUTH{!}!`. Further, if a delimiter comes immediately after the previous argument (or the macro), the argument it delimits is empty; in this case there is no need to use `{}` as in section 12.3.

Delimiters can serve to make the invocation of certain macros more intelligible. For example, the definition of `\buildrel` in plain T_EX starts with

$$\texttt{\textbackslash buildrel\#1\textbackslash over\#2\{...\}}$$

The `\over` merely indicates the end of the first argument: it is not expanded as a macro, so it won't produce a fraction bar. The second argument is undelimited. Likewise, the definition of `\bbuildrel` in section 11.14 said

$$\texttt{\textbackslash def\textbackslash bbuildrel\#1\_\#2\^{}\#3\{...\}}$$

Here again `^` and `_` are mere delimiters: they are not directly responsible for writing arguments 1 and 2 above and below argument 3. It's the `^` and `_` inside the replacement text that do that.

You can also have a "delimiter" before the first argument in a definition, or before the `{` if the macro has no arguments. The trick explained in section 12.2 is an example of this: after

$$\texttt{\textbackslash def\textbackslash twod/\{two-dimen\textbackslash -sional\}}$$

T_EX always expects to find the delimiter `/` immediately after the macro `\twod`. It's almost as if the macro name were `\twod/`, but there are two important differences:

- `\twod` can only have one meaning at a time, so the definition above erases any previous one. Even if you had previously said `\def\twod{2D}`, you're not allowed to use `\twod` by itself after the new definition:

        ! Use of \twod doesn't match its definition.

- \twod / works just the same as \twod/, because a space after a macro is ignored.

Here's another trap to be on guard against:

> \def\a$z{alpha and omega}
> \a $z=x+y$ Humpty Dumpty sat on a wall...

This will give the error message Missing $ inserted. and the output

> alpha and omega=x+y*HumptyDumptysatonawall*...

What went wrong? Remember, the delimiter $z is *not* part of the replacement text. TEX effectively replaces all of \a$z by alpha and omega, so the first $ never has a chance to do its stuff. It's only after the second $ that TEX goes into math mode.

For all these reasons, you should probably stay clear of delimiters without arguments, except in the simple case of abbreviations, or, occasionally, for a special need. Sure, you can define \1/2 so that it writes $\frac{1}{2}$ in your document. But isn't it just as simple to call your macro \half ?

Two error-recovery mechanisms

If you leave out by mistake the closing braces of a macro's argument (or its delimiter, if it's a delimited argument), TEX won't be able to figure out where it should end. Conceivably, it might read all the way to the end of your file still thinking it's inside the argument: that would most likely bust its memory and ruin the whole run.

To avoid this situation, TEX works on the assumption that arguments should never contain the \par control sequence, or its alias, a blank line. If TEX sees something like \toto{...\par, it assumes a mistake somewhere, issues a Runaway argument? meassage, and cuts its losses by stopping the expansion of \toto.

Of course, TEX gives you a way out. Define your macro with the construction

> \long\def\toto#1{...}

and its arguments will be under no restriction whatsoever. You should use this workaround carefully, since you're effectively giving TEX carte blanche to swallow hundreds of pages in one gulp...

Even with the \par -catching mechanism, mismatched braces are potentially catastrophic—imagine what happens if you have one left brace too many in a macro definition. As an additional strategy for error detection, TEX lets you declare a macro to be \outer, in the following sense: after a definition like

> \outer\def\toto{...}

\toto is treated even more strictly than \par in terms of where it can occur. Not only is it forbidden inside an argument—even the argument of a \long macro— but also inside definitions, alignment templates, and a few other places. The end of a file is subject to the same restrictions. A macro can be both \long and \outer.

12.5 Category codes

You've known for a long time that some of TEX's characters are special, like $, % , and so on. What makes them special? It turns out that the "meanings" of characters are not written in stone: if necessary, you can very well change them around.

Each of the 128 (or 256) characters that you can produce on your keyboard has a category in TEX, and it's the character's category that gives it its meaning. Here are the categories of all characters in plain TEX:

Category	Meaning	Characters
0	escape character	\
1	begin group	{
2	end group	}
3	begin/end math	$
4	alignment separator	&
5	end of line	CR
6	argument	#
7	superscript	^ , SUP
8	subscript	_ , SUB
9	ignored	NULL
10	space	SP, TAB
11	letter	A ... Z, a ... z
12	ordinary	all others
13	active character	~
14	comment	%
15	invalid	DEL

Here SUP, SUB, NULL and DEL stand for the (non-printable) characters with ASCII code 11, 1, 0 and 127—don't worry about them. TAB is the character you get by pressing the tab key on your keyboard; its ASCII code is 9, and it normally produces one or more spaces on your screen. (This is different from the "tab" character & of chapter 10.)

You don't have to memorize this table—you can always refer back to it when necessary. The important thing is that each special meaning is associated with a different category. (But it's good to know that letters have code 11 and other ordinary characters, like @, have code 12.) The only special meanings we haven't seen before are associated with categories 9 (ignored) and 15 (invalid). An ignored character is simply skipped over; an invalid character causes an error message.

You can change the category of a character at any time by saying

$$\texttt{\catcode`\}X\texttt{=}n$$

where X stands for the character, and n for the new category code. The curious `\ construction gives the *numerical value* of the following character; its use is essential, becuse TEX must have some way to know that the character is to be taken literally, and its special meaning (if any) disregarded.

As an application, suppose you have a financial report that has $'s all over. Rather than typing \$ every time, you can start your file with

$$\text{\catcode`\\\$=12}$$

From there on, $ is no longer the harbinger of mathematics: it has become an active character that prints a '$'! (So to make $1,000,000 you just have to type $1,000,000 —isn't that great?)

A \catcode assignment made inside a group is undone at the end of the group, so you can easily limit its reach. In the financial report, for instance, you might need the regular meaning of $ (begin/end math) to typeset some formulas. Easy enough: surround the formulas with

$$\text{{\catcode`\\\$=3 \$ ... \$}}$$

and after the group is closed, $ is again an ordinary character. But this isn't a very good solution if the formulas are interspersed with dollar amounts, since you'd have to switch back and forth several times. A better alternative is to start the file with \let\math$, before you first change the category of $. Then the original meaning of $ is preserved in the \math control sequence, and the formulas can be coded with \math ... \math, which admittedly looks funny, but is easy to type.

12.6 Active characters

A character of category 13, called an active character, is really a macro in disguise. So not all macros are control sequences, after all! Any character can be made active. In plain TEX only one character is active, namely ˜; as you know, it creates a unbreakable space. The commands that set things up that way are

$$\text{\catcode`\˜=\active   \def˜{\penalty10000 \ }}$$

Here \active is a control sequence that expands to 13, so you don't have to remember the code explicitly. Notice that changing the \catcode is just a preliminary, that makes ˜ into a macro; after that you still have to define the macro using \def.

You don't type a \ before an active character, either when defining it or when using it—if you do, you get a one-character control sequence, which is not at all the same. For example, \˜ and ˜ are quite different macros (see section 2.4).

A space is not discarded after an active character: if you type Dr.˜ Jekyll and Mr.˜ Hyde you'll get two spaces before each name, one from the ˜ and one explicit.

To deactivate a character, you can reassign its category code explicitly, presumably to whatever it was before the character was made active; for example, for ˜ you'd say \catcode`\˜=12. But if you mean the activation to be temporary to begin with, you should perform it inside a group: then it will go away when the group ends.

In section 9.8 we made * active to use it as an invisible digit. Here are some more situations where characters can profitably be made active:

● If you're typing German, you may prefer to generate an umlaut with a single keystroke, rather than using plain TEX's \" macro. No problem: just say

```
\catcode`\"=\active   \def"{\"}
```

and a " before a vowel will be enough to place an umlaut above it.

● The vertical bar | is made active in the `book.mac` file containing the macros for this book. Any material between vertical bars is printed verbatim, that is, exactly as it appears in the input file; backslashes, braces and so on are not interpreted as special characters. This is very useful when giving examples of TEX code. In addition, verbatim mode switches to a typewriter face. Here is, in essence, the relevant part of `book.mac`:

```
\def\makeordinary{\catcode`\&=12 \catcode`\{=12
  \catcode`\}=12 \catcode`\#=12 \catcode`\\=12 \catcode`\$=12
  \catcode`\_=12 \catcode`\^=12 \catcode`\%=12 \catcode`\~=12}
```

```
\catcode`\|=\active
\def|{\bgroup\makeordinary\obeylines\obeyspaces\tt%
  \def|{\egroup}}
```

When TEX sees a |, it starts by opening a group and making several special characters ordinary. It continues with \obeylines and \obeyspaces, so carriage returns and spaces are not combined (see below). Next it redefines |! The next time that a | is seen it indicates the end of verbatim mode, which is accomplished simply by closing the group. Everything, including the definition of |, reverts to its original state.

● In French typography it is conventional to leave some space before, as well as after, a colon or semicolon. But if you leave a space before a colon in your TEX file, you may get a line break there, and the colon at the beginning of a line, which is definitely wrong. In any case, beginning typists often have a hard time getting the spacing straight. So in a macro file for French typesetting, it's good to make the colon an active character and write a definition for it that takes care of all these details. Here's one possibility:

```
\catcode`\:=\active
\def:{\unskip~\string:\ \ignorespaces}
```

With this definition, a "wrong" input like `a:b` or `a :b` or `a: b` gives the same result as the right one, `a~: b`. In effect, TEX replaces the spaces that might precede or follow the colon with its own spaces. See sections 6.12 and 6.13 for \unskip and \ignorespaces.

In the definition of : you see the \string control sequence, which tells TEX to treat the following character as ordinary (unless the character is a space character: see the next paragraph). If \string weren't there we'd be in trouble when we got to a colon: TEX would first replace the : by its expansion; after executing \unskip~, it would again see a : and replace it by its expansion; and again, and again, until it ran out of memory. By temporarily making : into an ordinary character, we avoid this infinite recursion, the computer equivalent of perpetual motion.

`\string` can also be followed by a control sequence, in which case it generates the control sequence name, written in ordinary characters. Thus `\string\toto` prints "toto, regardless of whether `\toto` is defined, and of what its meaning might be. (See `\char` in the Dictionary to find out why a backslash prints as ".)

• Another important use of active characters is in the `\obeylines` macro (section 6.4). The basic idea is very simple: the carriage return is made active, and given the definition `\par`:

```
{\catcode'\^^M=\active \def^^M{\par}
 ... % lines are obeyed here
}      % return to normal
```

TEX reads the three-character combination `^^M` as if it were a carriage return. This trigraph is used instead of an actual carriage return whenever the focus is on the character itself, since a CR is invisible: it just causes a new line on your screen.

So far, so good; between the moment you make `^^M` active and the end of the group, TEX is obeying lines. Now let's try to define `\obeylines` to do the job of the first line above:

```
\def\obeylines{\catcode'\^^M=\active \def^^M{\par}}
{\obeylines ...
```

Something really wierd happens: TEX reads your whole file without doing a thing, and complains of a `Runaway definition?` at the end! What happened? Remember, with `\def`, TEX is learning the definition of a macro, not executing it. So when `\catcode'\^^M=\active` is read, no change takes place. TEX goes on to read `\def`, and arrives at the `^^M`. It's here that disaster strikes: `^^M` is still an end-of-line character, so TEX skips right on to the next line! The `^^M` never makes it into the definition, and the rest of the line is not even seen.

Even if we manage to put the `^^M` inside the definition, there's trouble later. Consider this new attempt:

```
{\catcode'\^^M=12%
   \gdef\obeylines{\catcode'\^^M=\active \def^^M{\par}}}
{\obeylines ...
```

The definition is made global because it occurs inside a group. The `%` at the end of the first line prevents the `^^M`, which is now an ordinary character, from being typeset. This time the definition is correctly read, but when it comes time to execute `\obeylines`, TEX turns up its nose:

```
! Missing control sequence inserted.
<inserted text>
                 \inaccessible
<to be read again>
                 ^^M
\obeylines ->\catcode '\^^M=\active \def ^^M
                                             {\par }
```

The `^^M` is still not recognized as an active character! The reason is that *the category used for a character in the replacement text of a macro is the one it had when the macro's definition was read*, no matter how many times it has been changed since. Here, then, ^^M is of category 12 when \obeylines is expanded, and so cannot be used after \def .

The solution, then, is to make ^^M active before the definition of \obeylines is read. And so we get essentially to the definition actually used by plain TeX:

```
{\catcode'\^^M=\active%
    \gdef\obeylines{\catcode'\^^M=\active \def^^M{\par}}}
```

(In fact, plain TeX says \let^^M=\par instead of \def^^M{\par} , but the difference isn't worth fussing about. If you really must know, see the Dictionary under \let .)

• The \obeyspaces macro, which is like \obeylines but changes the category of SP, is similar. The only difference is that it doesn't redefine SP every time it's called; rather, a meaning is assigned to the character once and for all, the meaning being the same as that of the macro \space , previously defined with \def\space{ } :

```
        \def\obeyspaces{\catcode'\ =\active}
        {\obeyspaces\global\let =\space}
```

While on the subject of \obeyspaces , even experienced users of TeX are often confused and dismayed by the fact that spaces at the beginning of a line don't seem to be obeyed. Remember, spaces have no effect in vertical mode! In order for spaces to be strictly obeyed, then, it is necessary to change the definition above to say

```
        {\obeyspaces\gdef {\leavevmode\space}}
```

12.7 How TeX reads and stores your text

When you read a text, your first task is to group the letters together into words, which is pretty easy, since there are spaces separating them. (It wasn't always so—look at any Greek papyrus. . .)

In the same way, TeX's first job when reading your input is to chop it up into "words," in a process called lexical analysis, which Knuth likens to chewing. Naturally, TeX's words aren't the same as ours—in fact, most of them consist of a single character. It's only when TeX sees the escape character \ and reads the subsequent control sequence name that it makes a word from more than one input character. Thus, the short text

```
        \kern 3pt $\alpha${\it code}\t@t@
```

is analyzed as follows: \kern, 3, p, t, SP, $, \alpha, $, {, \it, c, o, d, e, }, \t, @, t, @. Spaces after a control sequence made of letters don't form words, nor do they become part of the control sequence name: they're simply discarded, as we've seen before.

The way in which a line is chopped into words is affected by the categories of the characters on it. For example, when % is of category 14, as usual, characters after it on the same line are not even seen by TEX, and certainly not made into words. But if you make % an ordinary character, that's of course no longer the case.

What characters are escape characters is another critical factor in this mastication process. In section 12.1 we said that a control sequence is made of a \ followed by one or more letters or one non-letter. Now the truth is revealed: *any* escape character (one whose category code is 0) can introduce a control sequence, and the body is made up of one of more "generalized letters" (characters of category 11) or one generalized non-letter.

What this means is that if you set `\catcode'\@=0`, the input `\toto@toto0` will be interpreted as containing two control sequences and the character 0, rather than one control sequence and the characters @, t, o, t, o, 0, as it normally would. In fact, it's better to say that the input contains the same control sequence twice: after reading a control sequence, TEX doesn't remember what control character introduced it, only what characters form its name.

If you instead set `\catcode'\@=11`, you make @ a "letter" from TEX's point of view, so it can be part of control sequence names. In this case, `\toto@toto0` is analyzed into *one* control sequence `\toto@toto` and one character. We'll come back to this point later on.

Tokens

As we learned from our experiments with `\obeylines` in the preceding section, a character that is stored in a definition somehow carries with itself the category it had at the time of reading. To reflect this we will from now on speak of TEX's words as *tokens*. A token is either a character together with its category (which once assigned at reading time is never changed), or a control sequence. In terms of tokens, the input

```
\kern 3pt $\alpha${\it code}\t@t@
```

comes out as $\boxed{\text{kern}}$, 3_{12}, p_{11}, t_{11}, SP_{10}, $\$_3$, $\boxed{\text{alpha}}$, $\$_3$, $\{_1$, $\boxed{\text{it}}$, c_{11}, o_{11}, d_{11}, e_{11}, $\}_2$, $\boxed{\text{t}}$, $@_{12}$, t_{11}, $@_{12}$. Here we're writing character tokens with the category code as a subscript, and control sequence tokens inside a box. We do this to stress the indivisible character of tokens: once a control sequence has been read in, it's no longer thought of by TEX as made up of several characters.

Once read in, then, your input is entirely handled at the token level, no matter how many times it's shuffled around from macro to macro. *The replacement text of a macro is made up of tokens, and it never goes through the process of lexical analysis in TEX's "mouth" again.* (In other words, TEX is not a ruminant...)

This has an important application, the creation of protected macros. These are macros whose names include some character, say @, that is normally of category 12; of course, they must be defined at a time when @ is a "letter," that is, has category 11. Once @ is again of category 12, the macro can no longer be redefined, or even used directly—it is protected. But it will be encountered by TEX, and perform its function, if it occurs in the replacement text of other macros, which were also defined while @ was a letter.

Here's an example. Normally, saying `\t@t@` gives @̂t@, because the input `\t@t@` is divided into the tokens $\boxed{\tt t}$, @$_{12}$, t$_{11}$, @$_{12}$ (for the meaning of `\t` , see section 2.4). For the same reason, `\def\t@t@{TOTO}` would redefine `\t` as a macro with delimiters (section 12.4), rather than defining a new control sequence `\t@t@`. But if you say

`\catcode'\@=11`
`\def\t@t@{TOTO}` `\def\toto{\t@t@}`
`\catcode'\@=12`

you are actually defining `\t@t@`, and you are furthermore defining `\toto` to expand to the single token $\boxed{\tt t@t@}$—not to the string of characters `\t@t@`. So when you now type `\toto` you get the output TOTO, *even after* @ *is no longer a "letter!"* But you can't call `\t@t@` directly anymore, and you certainly can't clobber its definition by mistake. For a real-life example, see `\afterassignment` in the Dictionary.

12.8 Registers

In section 8.5 we saw that TEX has 256 slots in its memory to store boxes; they are called *box registers*. There are register classes for several other types of objects, each with registers numbered from 0 to 255:

Register Class	Type of Contents	Sample Assignment	Direct Usage	Indirect Usage
`\count`	integer	`\count3=17`	`\count3`	`\the\count3`
`\dimen`	dimension	`\dimen0=.3in`	`\dimen0`	`\the\dimen0`
`\skip`	glue	`\skip5=2pt minus 1pt`	`\skip5`	`\the\skip5`
`\muskip`	math glue	`\muskip4=5mu plus2mu`	`\muskip4`	`\the\muskip4`
`\toks`	token list	`\toks3={toto\hfil}`	`\toks3`	`\the\toks3`
`\box`	box	`\setbox9=\hbox...`	see below	not available

Some comments about the second column:

• Integer registers are straightforward; all you might (or might not) want to know is that the largest integer they'll hold is 21474483647, or $2^{31} - 1$, and the smallest one is -21474483647.

• Dimension registers hold dimensions whose absolute value is less than 16384 pt, or 18.892 feet, or 5.7583 meters. Dimensions are converted to a minute unit, called the scaled point (sp), and rounded to the nearest unit. There are $2^{16} = 65536$ scaled points in a point.

• A `\skip` register holds glue specification, which consists of three dimension components: the natural component, the stretchability, and the shrinkability (section 5.3). The last two can be infinite, as in the case of springs.

• A `\muskip` register holds math glue, which is glue specified in math units (see the end of section 11.7).

• A \toks register holds a list of tokens, which is somewhat like a macro without arguments, but more efficient for certain operations. We saw such lists in action in section 7.2.

• Finally, a \box register holds, surprisingly enough, a box. The storage and retrieval of boxes were presented in section 8.5, which you're urged to reread at this point; they differ from the corresponding operations for other registers. In particular, the left-hand side of a box assignment says \setbox9, rather than \box9. The latter command uses the box and empties the register. The other commands to use a box are \copy, \unhbox, \unvbox, \unhcopy, \unvcopy and \vsplit. All were covered in section 8.5, except for the last, which was explained in section 8.12.

Storing something in a register

In all other cases except boxes, the command that does the storing, or assignment, is the same: the register name, followed by an = sign (optional), followed by an object of the appropriate type. The object can be specified from scratch, as in the table, or make reference to other registers. We'll discuss this second possibility in detail below. A token list specified from scratch must come within braces; macros are not expanded while the list is being read and stored.

All assignments should be preceded by \global if their effect is to last beyond the end of the current group.

Our discussion of registers should also include all of TEX's special variables like \parindent, \baselineskip, and so on. Such a variable is essentially a register of one of the first five types above, having a special name and a special effect on TEX's actions. The variable can be used wherever a register of the same type can.

Naming a register

For all the reasons mentioned in section 8.5, it's not a good idea to used register numbers explicitly, except for those that plain TEX specifically designates as scratch registers, and even those only briefly. For any other use, you should request a named register, using one of the commands \newcount, \newdimen, \newskip, \newmuskip, \newtoks and \newbox.

The way you use these allocation commands is very simple. After you say, for instance,

\newcount\mycount

the control sequence \mycount becomes synonymous with \count n, for some n that's not associated with any other counter so defined. So if everyone abides by this discipline, you can be sure that \mycount won't be overwritten by somebody else's macro. But \count n is still a valid way to access the same register, so if you, or anybody else, start using register numbers at random, \mycount will be at risk. For the allocation system to work everyone has to cooperate.

The \newbox command, as we've seen, is slightly different: \newbox\mybox makes \mybox equal to a number, not a box. To refer to the box you write \mybox after \box, \setbox, \copy, and so on.

The registers that are safe for temporary use are:

```
\count255
\dimen0, ..., \dimen9, \dimen255
\skip0, ..., \skip9, \skip255
\muskip0, ..., \muskip9, \muskip255
\toks0, ..., \toks9, \toks255
\box0, ..., \box9
```

Never use other registers by number, unless you know exactly what you're doing. And if you do know what you're doing, you won't use other registers by number.

You break this convention at your own risk. You'll find yourself wondering why the macros that were working yesterday aren't working today.

Inspecting a register

You can inspect at any time the contents of a register by writing `\showthe` followed by the register name. When TEX encounters that instruction, it stops and shows the information on your screen; in order to get it started again you must type CR. The information is also saved in the log file.

You can't use `\showthe` with boxes, but you can instead say `\showbox`, followed by the box number. This will write the contents of the box (in symbolic form) into the log file; to get them on the screen as well you must set `\tracingonline=1` (compare section 3.3).

Using a register

Registers other than boxes can be used in two ways, indicated in the two rightmost columns of the table. The difference is subtle, yet fundamental.

Suppose you've set, say, `\mycount=1990`, where `\mycount` was defined with `\newcount`. (Remember that `\mycount` is the same as `\count` n, for some n that should remain unknown.) If, at some later time, TEX encounters `\mycount` by itself, it assumes that you're about to assign another value to the register, *unless it has reason to expect an integer quantity at this point.* If it is expecting an integer, its expectations are satisfied: the integer is 1990. TEX has used the contents of `\mycount` directly.

But if TEX encounters `\the\mycount`, it replaces `\the\mycount` by 1990, and carries on: these four tokens will be processed as if you'd typed them at that point. Perhaps TEX was expecting an integer here too: in that case it will consider 1990 as the integer's first four digits, and read on to see if there are more. The important point is that `\the` generates a string of tokens which blend with the preceding and following tokens; but `\mycount` by itself generates no tokens, rather the register's contents are treated as an abstract object of a certain type (here an integer) for which there is a pressing need.

To make these ideas a bit firmer, here's another example. After the `\vskip` primitive, TEX expects to see the specification of some glue. The specification might be 2pt plus 1pt, for example. Or it might be `\skip5`; if this register had been previously set with `\skip5=2pt plus 1pt`, the effect would be the same.

Or the specification might be \the\skip5; TEX would replace \the\skip5 with 2pt plus 1pt, and the effect again would be the same. Wait—not quite: \vskip\the\skip5 minus 1pt causes a skip of 2pt plus 1pt minus 1pt, but \vskip\skip5 minus 1pt causes a skip of 2pt plus 1pt, and a new paragraph starting with minus 1pt!

To summarize, naming a register by itself can mean either that you're about to assign a new value to it, or that you're using its value directly. The latter only makes sense if TEX is expecting to see an object of the corresponding type. The important question, then, is: At what times is TEX expecting an integer (or a dimension, or a box, etc.)?

Uses of integers

Here are some of the most common situations where an integer is expected:

• After \count, \dimen, and other register class names; after \box, \copy and friends; and after \ht, \dp, \wd. So you can say \dimen\mycount=10pt; if \mycount had been given the value 188, this sets \dimen188 to 10 pt.

• In an assignment to a \count register, or to any of TEX's integer variables, which are legion (pages 272–273 of *The TEXbook*). Thus you can say \count255=\mycount, or \hangafter=\count255, or \pageno=\hangafter (\pageno is the current page, and happens to be the same as \count0).

• After \number and \romannumeral, which return the decimal representation and the roman numeral representation of the integer.

• After the tests \ifodd and \ifeven (see section 12.9). After \ifnum two integers are expected, separated by <, = or >.

When TEX is expecting an integer it will accept the contents of a dimension register (used directly). In that case it expresses the register contents in scaled points, the units in which the dimension is stored. Thus, \dimen1=1pt followed by \mycount=\dimen1 gives \mycount the value 65536. But \the\dimen1 expands to 1.0pt, so \mycount=\the\dimen1 gives \mycount the value 1, and prints .0pt.

Uses of dimensions

The most common times when a dimension is expected are:

• In an assignment to a \dimen register, or to any of the dimensions of a box register, or to any of TEX's dimension variables, like \parindent, \hsize and \vsize (page 274 of *The TEXbook*). Examples: \dimen1=6pt, \ht1=\dimen1, \wd1=\hsize.

• After \kern, \raise, \lower, \moveleft, and \moveright.

• In place of the ellipses in the constructions \hbox to ... and its relatives, \vrule height ... and its relatives, and so on.

• Whenever a glue specification is expected (see below). That is, the glue can be specified by means of its components: \skip1=\dimen0 plus \ht0 minus \parindent.

- After the `\ifdim` test (section 12.9) two dimensions are expected, separated by `<`, `=` or `>`.

Whenever a dimension is expected, TₑX will also accept the contents of a `\skip` register (used directly), and discard its stretch and shrink components. For example, plain TₑX sets `\medskipamount` to be 6pt plus 2pt minus 2pt; saying `\kern\medskipamount` is the same as saying `\kern 6pt`.

An integer preceding a unit also gives a dimension: `\dimen1=\mycount pt`. Another important way to specify a dimension is by multiplying an existing dimension by a factor: `\kern -.5\dimen7`. This type of specification is not available for integers, although a minus sign is allowed: `\mycount=-\pageno`. It's not available for glue either: if you say `\skip3=-.5\skip0` TₑX will throw out the stretch and shrink components of `\skip0` before performing the multiplication.

Uses of glue

And here are the most common times when a glue specification is expected:

- In an assignment to a `\skip` register, or to any of the dimensions of a box register, or to any of TₑX's glue variables, like `\parskip` and `\baselineskip` (page 274 of *The TₑXbook*). For example, `\skip1=6pt plus 2pt minus 2pt`, `\parskip=\skip1`.

- After `\hskip` and `\vskip`.

Uses of math glue

And the only times when a math glue specification is called for are:

- In an assignment to a `\muskip` register, or to one of the math glue variables `\thinmuskip`, `\medmuskip` and `\thickmuskip` (page 274 of *The TₑXbook*).

- After `\mskip` and `\mkern` (in the second case a math dimension is all that's needed).

In both cases nothing else will do: TₑX won't convert from normal glue to math glue, or vice versa.

Uses of token lists

TₑX expects to see a token list in an assignment to a `\toks` register, or to one of its token list variables (page 275 of *The TₑXbook*). When a token list is first read in, the macros in it are not expanded, and braces must surround the list: `\toks0={...}`. The macros will be expanded when the list is used, by preceding the register name with `\the`. Token lists are almost always used indirectly, that is, together with `\the`: the only use for `\toks0` by itself is on either side of an assignment.

Arithmetic on registers

You must be wondering if you can do arithmetic operations on the numbers and dimensions you store in registers. You can; but it's not a pretty sight. TₑX is not a general-purpose programming language, and the need to make it absolutely device-independent restricts the arithmetic to what can be done relatively easily with integers.

To add to or subtract from a `\count`, `\dimen`, `\skip` or `\muskip` register, you use the `\advance` command:

```
\advance\pageno by 1     \advance\skip0 by 0pt plus 1fil
\advance\dimen1 by -3pt \advance\muskip5 by-.5\thinmuskip
```

You can have after `by` anything that you can have on the right-hand side of an assignment to the same class of registers. Notice that there is no backslash before `by`.

Multiplication and division are also allowed, by only by integers:

```
\divide\mycount by 3     \multiply\parindent by 2
\multiply\dimen1 by -6  \divide\muskip5 by\count255
```

What would be written in Pascal as `\count2:=3+0.5*\count1` comes out as

```
\count2=\count1 \divide\count1 by 2\advance\count1 by 3
```

It's a good thing there isn't a whole lot of arithmetic to do!

Variables like `\parindent`, `\thinmuskip` and `\hangafter` can also be modified with `\advance`, `\multiply` and `\divide`. But some other things that you can assign values to, such as the dimensions of a box, are somehow left out: to decrease the height of `\box1` by 10 pt, you must say

```
\dimen0=\ht1 \advance\dimen0 by -10pt \ht1=\dimen0
```

However, `\ht1=.5\ht1` works fine, since it's an assignment.

12.9 Conditionals

Like all programming languages, T_EX possesses conditionals, constructions that choose one or another course of action depending on the current value of certain variables. Using conditionals it is possible to build up other control structures, such as loops to iterate one or more commands automatically. This section won't cover general control structures, or even all the uses of conditionals; after all, this book is supposed to be an introduction to T_EX only.

In the `fancy.tex` file of section 7.2 a part of the `\headline` token list says

```
\ifodd\pageno\the\oddpagehead\else\the\evenpagehead\fi
```

When T_EX reads `\ifodd`, it looks for an integer after that, as explained in the previous section. Here the integer is supplied directly from a register, but it could also be written explicitly, or come from the expansion of a `\the`, or whatever. If the number is odd, T_EX continues reading and doing its stuff till it reaches `\else`; it then skips everything till the next `\fi`. Here the result would be to read `\the\oddpagehead`, so the `\oddpagehead` token list would be used at this point. If the number is even, contrariwise, T_EX skips the text till the `\else`, but reads what follows till the `\fi`, so it's `\evenpagehead` that would be used.

Some warnings

All of this probably seems obvious to you, which is why we used this code unapologetically in section 7.2. But there are some aspects of T_EX's conditionals that may seem counterintuitive, especially if you're accustomed to other programming languages, so it's good to go over them briefly.

• A TEX conditional chooses between two *texts,* which don't have to be actions or commands. Thus you can say

```
\parindent=\ifodd\pageno 20 \else 10 \fi pt
```

rather than `\ifodd\pageno\parindent=20pt\else\parindent=10pt\fi` as in some other programming languages.

• Each of the two texts can have unbalanced braces; the important thing is that the overall text *after* a choice is made be balanced. If you say

```
\ifodd\pageno \toto{\toto \else \otot{\otot \fi }
```

TEX will see `\toto{\toto}` if the page number is odd, and `\otot{\otot}` if it's even, so thing come out right either way. Remember, TEX doesn't pay any attention to the part of a conditional that it's skipping over.

• Once TEX starts evaluating a condition, it's committed to it. So you'd better make sure this doesn't happen while something else is underway. The following code has stumped countless aspiring wizards:

```
\pageno=0\advance\pageno by1\ifodd\pageno ODD \else EVEN\fi
```

Oddly, TEX prints EVEN. Do you see what happened? TEX sets `\pageno` to 0, then is told to increment it by 1. But wait—maybe 1 is just the first digit of the increment! TEX has to read ahead to see where the number ends, so it evaluates the `\ifodd`. At that time, `\pageno` is still 0! What TEX sees next is EVEN. It decides that's not part of the number, so it increases `\pageno` by 1, but doesn't do the test again; the whole conditional has already been effectively replaced by its `\else` portion.

Fortunately this sort of thing can be easily avoided by always leaving a space after a number. The space is absorbed when TEX reads the number, and doesn't show in the output. We made tacit use of this fact in many examples throughout the book.

OK, so what's the value of `\pageno` after

```
\pageno=0\advance\pageno by1\ifodd\pageno 0 \else 1\fi
```

and what (if anything) does TEX print?

Other conditionals

TEX has many other tests besides `\ifodd`, of which we'll only talk about some. In almost all cases the construction is the same:

$$\textit{test text1} \ \texttt{\textbackslash else} \ \textit{text2} \ \texttt{\textbackslash fi}$$

Either *text1* or *text2* can be empty (or even both, but then there isn't much point to the conditional). If *text2* is empty, the `\else` is not necessary.

Here then is a (non-exhaustive) list of TEX's conditionals:

• `\ifnum` *integer1 relation integer2*; the relation is either < , = or > . A simple footline macro for a technical report format might say, for instance,

```
\def\footline{\hfil \ifnum\pageno=1\else\folio\hfil}
```

Notice that in this case *text1* is empty.

• \ifodd *integer*. There's no \ifeven, but get the same effect by switching the *text1* and *text2*.

• \ifdim *dimension1 relation dimension2*; the relation is either <, = or >. A macro to select the wider of two boxes can be written

```
\def\pickwider#1#2{\ifdim\wd#1>\wd#2\box#1\else\box#2\fi
```

• \ifmmode is true if T_EX is in math mode (text or display). Useful if you want abbreviations that should work both inside and outside of math mode:

```
\def\a{\ifmmode\alpha\else$\alpha$\fi}
```

• \ifvoid *box number* is true if the corresponding box is undefined. Useful to avoid the error of trying to use an undefined box: \ifvoid1\else\box1\fi.

• \ifcase *integer* is the only conditional with a different syntax, because it's capable of choosing between $n + 1$ actions, depending on whether *integer* has the value $0, 1, \ldots, n$. The use of \ifcase is easier to learn by example than from an explanation:

```
\def\monthname{\ifcase\month\or January\or February\or
    March\or April\or May\or June\or July\or August\or
    September\or October\or November\or December\fi}
```

Custom-made tests

In addition to all the tests above, you can create new tests at any time using plain T_EX's \newif command. In section 7.2 we said

```
\newif\iftitlepage \titlepagetrue
    ...
\footline={\iftitlepage\the\titlepagefoot
    \global\titlepagefalse
    \else\ifodd\pageno\the\oddpagefoot
        \else\the\evenpagefoot\fi\fi}
```

Saying \newif\iftitlepage defines three control sequences: \iftitlepage, \titlepagetrue and \titlepagefalse. After that, saying \titlepagetrue makes \iftitlepage test true, and saying \titlepagefalse makes it test false. Inside \footline, we test if the current page is a title page or not; if it is, we use the special footline \titlepagefoot, and say \global\titlepagefalse so the next page will no longer be a title page. We have to say \global because the effects of \titlepagetrue and \titlepagefalse are local to the current group.

This example shows also that conditionals can be nested: the \else portion of the \iftitlepage conditional contained itself a complete conditional. We said before that T_EX ignores braces and other groups when it's skipping over the rejected portion of a conditional; but it does keep an eye out for \if...\fi pairs, so it will only stop at the correct \fi.

12.10 For the aspiring wizard

To wrap up this chapter we will try to show how you can make use of Appendix B of *The TEXbook*, even if you don't understand its details in full. We have already created some macros, like `\itemitemitem` and `\bbuildrel`, by mimicking plain TEX—here we'll do it wholesale and shamelessly.

The `\cases` macro

This is one of the simplest alignment-making macros in plain TEX, so we start by trying to understand how it works, based on TEX's primitives. Its definition appears on page 362 of *The TEXbook*:

```
\def\cases#1{\left\{\,\vcenter{\normalbaselines\m@th
    \ialign{$##\hfil$&\quad##\hfil\crcr#1\crcr}}\right.}
```

First we get rid of the background noise. The `\left\{...\right.` know already: it creates left braces the size of the alignment. The alignment is placed in a `\vcenter`, inside which the environment is normalized: `\normalbaselines` copies the interline spacing information from `\normallineskip` to `\lineskip`, and so on (which explains why `\openup` won't work inside `\matrix`). Moreover, `\m@th` makes sure that `\mathsurround` is zero: this variable, of which nothing had been said so far, is the amount of horizontal glue that TEX puts between a formula and the surrounding text.

The alignment proper starts with `\ialign`, which is just `\halign` with yet another initialization, `\tabskip=0pt` (better safe than sorry). So we're really dealing with the following alignment:

```
\halign{$#\hfil$&\quad#\hfil\crcr#1\crcr}
```

As we saw in section 12.4, a double sharp `##` encountered when a macro is read in turns into a single `#` when the macro is executed. The only thing that's unfamiliar here is the `\crcr` control sequence, a primitive that turns into `\cr` unless it's already placed right after a `\cr` or `\noalign{...}`. We'll see it in action in a minute.

Suppose now that `\cases` is used in the following way:

```
\cases{A & if $x=1$, \cr
       B & otherwise.\cr}
```

The argument to `\cases` gets plugged into the replacement text in place of `#1`, so what TEX ends up seeing is

```
\halign{$#\hfil$&\quad#\hfil\crcr
       A & if $x=1$, \cr
       B & otherwise.\cr\crcr}
```

The first `\crcr` turns into `\cr`, indicating the end of the preamble. According to the preamble, the first entry of each column is read in math mode, the second isn't.

The second `\crcr`, at the end of the alignment, is superfluous, because the argument ended with `\cr`. This use of `\crcr` just makes `\cases` a little bit user-friendlier, making up for a missing `\cr` at the end of the argument.

A new macro: \Eqalign

One sometimes wants to arrange to formulas like this:

$$u_2 + v_2 = x^2 + y^2 \qquad u_2' + v_2' = x^4 + y^4 \qquad u_2'' + v_2'' = x^3 + y^3$$
$$v_3 = x^3 + y^3 \qquad v_3' = x^3 + y^3 \qquad v_3'' = x^4 + y^4$$
$$v_3 = x^4 + y^4 \qquad v_3' = x^2 + y^2$$

A naïve solution is to place three \eqalign s side by side:

```
$$\eqalign{...}\quad\eqalign{...}\quad\eqalign{...}$$
```

That works as long as all the entries have the same height, but is inadequate in general, since it doesn't guarantee that corresponding rows of the three alignments match. Can we generalize \eqalign so it takes several pairs of entries per row? The new macro should be used somewhat like \matrix:

```
$$\Eqalign{
  u_2+v_2&=x^2+y^2&u'_2+v'_2&=x^4+y^4&u''_2+v''_2&=x^3+y^3\cr
  v_3    &=x^3+y^3&v'_3     &=x^3+y^3&v''_3      &=x^4+y^4\cr
  v_3    &=x^4+y^4&v'_3     &=x^2+y^2                     \cr
}$$
```

Let's inspect the definition of \eqalign on page 362 of *The T_EXbook*:

```
\def\eqalign#1{\null\,\vcenter{\openup\jot\m@th
  \ialign{\strut\hfil$\displaystyle{##}$&
    $\displaystyle{{}##}$\hfil\crcr
  #1\crcr}}\,}
```

We see, incidentally, why \eqalign doesn't respond to springs in its entries: when an entry with \hfill is plugged into the first template, the resulting text is of the form

```
\hfil$\displaystyle{...\hfill}$
```

There is a pair of braces between the dollar signs. Unlike the situation in horizontal and vertical modes, groups in math mode create boxes for the subformulas they enclose. Any spring in a subbox is totally powerless in the enclosing box. To understand this point better, run the following experiments, and explain the results:

```
\hbox to \hsize{a\hfil{b\hfill c}}
\hbox to \hsize{a\hfil\hbox{b\hfill c}}
\hbox to \hsize{a\hfil$b\hfill c$}
\hbox to \hsize{a\hfil${b\hfill c}$}
\hbox to \hsize{a\hfil{$b\hfill c$}}
```

We also notice an empty group {} just before the # in the second template. This ensures that the operator with which the second half of each row normally starts gets the appropriate amount of spacing around it. For more details, see the end of section 11.18.

But we're getting sidetracked. To make our `\Eqalign` macro, we make the preamble periodic, by copying over the existing two templates after a double ampersand:

```
\catcode'\@=11
\def\eqalign#1{\null\,\vcenter{\openup\jot\m@th
   \ialign{\strut\hfil$\displaystyle{##}$&
     $\displaystyle{{}##}$\hfil&&
     \qquad\hfil$\displaystyle{##}$&
     $\displaystyle{{}##}$\hfil\crcr
  #1\crcr}}\,}
\catcode'\@=12
```

As you can see, you don't have to be a wizard to get a lot of mileage out of existing macros!

Systems of equations

Our next goal is to typeset the following system of equations:

$$\left\{ \begin{array}{l} 2x \quad + \quad 3y \quad + \quad 4z \;\; = a_1 + b_1 + c_1 \\ 22x^2 \;\; - \;\; 33y^2 \;\; + \;\; 44z^2 = a_2^2 + b_2^2 \\ 222x^{11} + 333y^{11} - \quad 4z \;\; = a_3 \\ -7x \quad - \quad 36y \quad + 478z^3 = b_4 \end{array} \right.$$

One idea is to use `\matrix`, as on page 158:

$$\left\{ \begin{array}{rlcrlcrlcl} 2 & x & + & 3 & y & + & 4 & z & = & a_1 + b_1 + c_1 \\ 22 & x^2 & - & 33 & y^2 & + & 44 & z^2 & = & a_2^2 + b_2^2 \\ 222 & x^{11} & + & 333 & y^{11} & - & 4 & z & = & a_3 \\ -7 & x & - & 36 & y & + & 478 & z^3 & = & b_4 \end{array} \right.$$

But this puts too much spacing between columns and not enough between rows, and it also forces us to type monstrosities like

```
\hfill 22&x^2\hfill &-&\hfill 33 &y^2\hfill
  &+&\hfill 44&z^2\hfill&=&a_2^2+b_2^2\hfill\cr
```

to format the rows. Instead, let's try to adapt `\eqalign` again, this time with a preamble of the form

               & *coefficient* & *variable* & *operator* `\cr`

The coefficients will be pushed right and the variables pushed left by springs. To get the right amount of spacing around the operators, we surround the corresponding `##` with empty groups. Also, we may as well throw in the left braces. The macro definition comes out fairly simple:

```
\catcode'\@=11
\def\system#1{\left\{\vcenter{\openup1\jot\m@th
   \ialign{&\hfil$##$&$##$\hfil&\strut${}##{}$\crcr
  #1\crcr}}\right.}
\catcode'\@=12
```

The code for the actual system, too, is quite natural. Notice that the last column doesn't have coefficients:

```
$$\system{
  2&x        &+&3  &y      &+&  4&z    &=&&a_1+b_1+c_1\cr
 22&x^2      &-&33 &y^2     &+&  44&z^2&=&&a_2^2+b_2^2\cr
222&x^{11}&+&333&y^{11}&-&  4&z    &=&&a_3         \cr
 -7&x        &-&36 &y      &+&478&z^3&=&&b_4         \cr
}$$
```

Numbering several equations

To number the formulas in a \displaylines, one must go through contortions with \hfill, \llap and \rlap:

```
$$\displaylines{
  \rlap{(3)}\hfill ....... \hfill            \cr
                  \hfill ....... \hfill\llap{(4)}\cr
}$$
```

This is not only a nuisance, but also a source of errors. Let's try to create macros \displaylinesno and \ldisplaylinesno to be used like \eqalignno and \leqalignno.

We start from the definition of \displaylines, also on page 362 of *The TEXbook*:

```
\def\displaylines#1{\displ@y\halign{
  \hbox to\displaywidth{$\@lign\hfil\displaystyle##\hfil$}
  \crcr#1\crcr}}
```

The single template is essentially a box of full width \displaywidth, which is the analogue of \hsize inside displays. It's straightforward to add a column at the right with zero width, using \llap:

```
\catcode`\@=11
\def\displaylinesno#1{\displ@y\halign{
  \hbox to\displaywidth{$\@lign\hfil\displaystyle##\hfil$}&
    \llap{$##$}\crcr
  #1\crcr}}
\catcode`\@=12
```

For labels on the left we need to be a bit craftier, since we want to keep the syntax of \leqno and \eqalignno, which specifies the label after the equation. We can still make a box of zero width, this time with \rlap, but we have to move it all the way across the display with appropriate kerns:

```
\catcode`\@=11
\def\ldisplaylinesno#1{\displ@y\halign{
  \hbox to\displaywidth{$\@lign\hfil\displaystyle##\hfil$}&
    \kern-\displaywidth\rlap{$##$}\kern\displaywidth\crcr
  #1\crcr}}
\catcode`\@=12
```

The definition of `\eqalignno` and `\leqalignno` in plain TEX (still on the same page of *The TEXbook*) is more complicated, because it cannot rely on an entry of full width. It achieves centering by playing with the `\tabskip` variable, discussed in section 9.12. We don't have to worry about this problem here, but if you feel adventurous you should try to disect those macros.

Here's an example of `\ldisplaylinesno` in action:

$$(1) \qquad\qquad \sin(a+b) = \sin a \cos b + \cos a \sin b$$

$$(1') \qquad\qquad\qquad \sin(2x) = 2 \sin x \cos x$$

```
$$\ldisplaylinesno{
  \sin(a+b)=\sin a\cos b+\cos a\sin b &(1) \cr
  \sin(2x)=2\sin x\cos x              &(1')\cr
}$$
```

13
Dictionary and Index

There you are, now—a savvy user of TEX, with a firm grasp of all the basic features. Should you be so inclined, you'll have no trouble at this point reading even the small-print sections of *The TEXbook*, and making your way into the select rank of TEX wizards. Or you can relax and enjoy the scenery—the knowledge you have already acquired will be sufficient to typeset just about any document.

Meanwhile, we hope this book will continue to be of use. The following Dictionary and Index contains all the control sequences discussed in the past twelve chapters, and adds some new ones. It also contains the main concepts that we've discussed, with references to the appropriate commands.

We have deliberately repeated information and suggestions from the "textbook" chapters, to make this chapter reasonably self-contained. But we also have made liberal use of cross-references for those who would reread the relevant sections.

Most control sequences listed here are primitives or macros from plain TEX. Conversely, most primitives and plain TEX macros are here, but we didn't include those that only a wizard might need. Macros that don't belong to plain TEX are indicated as such.

Backslashes introducing control sequences have been ignored in alphabetizing. Non-alphabetic characters are given in their ASCII order, which is the following:

> ! " # $ % & ' ( ) * + , - . / : ; < = > ? @ [\ ] ^ _ ` { | } ~

The "non-printable" characters represented by CR, DEL, NULL, SP, SUB, SUP, TAB are indexed under these abbreviations.

!	Produces an exclamation mark in text and in math mode. See also `\spacefactor` .
\!	Page 138. Math mode only. Produces a negative thin space, that is, brings the adjacent symbols closer together by 3 math units:
	`$$\int\!\!\!\int_{\cal D}f(u,v)\,du\,dv$$` .... $\iint_{\cal D} f(u,v)\,du\,dv$
!'	Page 19. Produces the Spanish ¡.
"	Pages 18, 36, 174. Produces right double quotes ", like '' . Also, when T$_E$X is expecting to read an integer (page 181), " announces that the number is written in base 16: see integers. To make " stand for an umlaut, see page 174.
\"	Pages 20–21, 174. Places an umlaut, or dieresis, over the following character: ü. Works in text mode only; for math mode, see `\ddot` . For ï you must type `\"\i` rather than `\"i` . See also the preceding entry.
#	Pages 17–18, 103, 167, 172. Used in the preamble of an `\halign` or `\valign` to indicate where an entry should be plugged in (pages 103, 119). Many examples were given in chapter 9. Also used, together with a digit, in the definition of a macro, to indicate where an argument should be inserted: see page 167, and examples on pages 51, 62, 92, etc.
\#	Page 17. Produces a sharp, or hash mark #. See also `\sharp` ♯.
##	Pages 169, 186. Used to represent a # in the replacement text of a macro.
$	Pages 15, 17, 22, 25–26, 81, 130–131, 172. Used to go in and out of text math mode. To neutralize this special meaning, see page 173.
\$	Page 17. Produces a dollar sign $.
$$	Pages 22, 25, 82, 130. Used to go in and out of display math mode. Also useful in centering non-math material: see examples on pages 90, 105, etc.
%	Pages 17–18, 172. Introduces a comment, or text disregarded by T$_E$X. Its effect extends to the end of the current line, including the carriage return that terminates it. Often necessary inside macro definitions to prevent spurious spaces: see examples on pages 35, 97, 175, etc.
\%	Page 17. Produces a percent sign %.
&	Pages 17–18, 103, 119, 122. Used to separate entries in the same row of a horizontal alignment obtained with `\halign` or `\+` , or entries in the same column of a `\valign` . Special cases are discussed on pages 106–107. Also used with many macros that perform alignments: see pages 147, 154–158, 186–190.
	Another, unrelated use of & is explained under `\dump` .
\&	Page 17. Produces an ampersand &. Page 125 suggests a redefinition useful if you have many alignments made with `\+` .
&&	Page 106. In the preamble of an alignment, means that the following templates are to be repeated cyclically. See examples on pages 111, 188.
'	Pages 18, 36, 137. Produces right single quotes ' in text, and a prime ′ in math mode. See also quotes and `\spacefactor` .
	When T$_E$X is expecting to read an integer (page 181), ' announces that the number is written in base eight: see integers.
\'	Page 20. Places an acute accent over the following character: é. Works in text mode only; for math mode, see `\acute` . For í you must type `\'\i` rather than `\'i` . The discussion about " on page 174 is relevant if your text has many accents.

, ,
: Pages 18–19, 36, 137. Produces right double quotes " in text, and a double prime " in math mode. See also quotes.

(
: Produces a left parenthesis in text and in math mode. As a math delimiter, it can grow arbitrarily large with `\left` (pages 146–147).

)
: Produces a right parenthesis in text and in math mode. As a math delimiter, it can grow arbitrarily large with `\right` (pages 146–147). See also `\spacefactor`.

*
: Pages 61, 133. Produces an asterisk in text * and in math mode ∗. The latter is synonymous with `\ast` and is a binary operator. Redefined on page 111 to create spacing equal to the width of a digit.

TeX types a * on your screen when it expects more input: page 16. At the beginning of an interactive run (no file name give) the prompt is ** instead. This means that TeX is ready to read a file name: if you type `Hello` it will look for a file called `Hello.tex`, rather than printing Hello. So if you actually want to type text at the terminal, you must start with a control sequence, such as `\par` or `\relax`, which are inoffensive.

The ** prompt has another function: it says that at that point, and only at that point, TeX can read an encoded format file. For details, see `\dump`.

*
: Indicates an allowed break at a multiplication in a mathematical formula; if the break is realized, a `\times` is inserted. For example, `$(x+y)\*(z+t)$` will come out as $(x+y) \times (z+t)$ if it has to be broken across lines, but as $(x+y)(z+t)$ otherwise.

+
: Pages 19, 133. Produces a + in text and in math mode; in the latter case it is treated as a binary operator. On page 149 it was used to construct a complicated new symbol.

\+
: Pages 122 and following. Starts a tabulated line, which should end with `\cr`. In the line tabs can be set and used with `&`, or deleted with `\cleartabs`.

This macro is `\outer`, that is, it is not allowed to appear in macro definitions and in certain other situations. To get around this, sue `\tabalign`, which is otherwise entirely equivalent to `\+`.

±
: See `\pm`.

,
: Produces a comma in text and in math mode; in the latter case it's automatically followed by a thin space: (a, b). If you use it to separate groups of digits in a large number, this spacing is undesirable: code `$75{,}000$` to get rid of it. See also `\spacefactor`.

\,
: Pages 138, 140. Math mode only. Leaves a thin space ∣ equal to 3 math units. Normally this space is inserted at the appropriate places automatically; the most common case where it should be explicitly used is before differentials, as in $\int x^2 \, dx$ from `$\int x^2\,dx$`. See also `\!`.

-
: Pages 19, 133. Produces a hyphen - in text and a minus sign − in math mode; in the latter case it is treated as a binary operator. On page 150 it was used to construct a complicated new symbol, but this is best done with `\relbar`. See also `\hyphenation`.

\-
: Page 166. Discretionary hyphen: tells TeX where a word can be broken between lines. Useful when TeX's automatic hyphenation process fails, as it does occasionally. For instance, TeX doesn't know how to hyphenate "manuscript," so if you happen to have an overfull line ending with that word you can help TeX by writing `man\-u\-script`. Also, TeX won't hyphenate compound words or words starting with a capital; you can override that with discretionary hyphens. See also discretionary, `\hyphenation` and `\showhyphens`.

\-\-
: Page 19. Produces an en-dash – in text, except with typewriter fonts.

`---`	Page 19. Produces an em-dash — in text, except with typewriter fonts.
∓	See `\mp`.
.	Produces a period in text and in math mode (page 130). Also used in math mode with `\left` and `\right` to stand of a dummy delimiter: see page 147 and `\abovewithdelims`. See also `\spacefactor`.
`\.`	Page 20. Places a dot above the following character: ȧ. Works only in text mode; for math mode, see `\dot`.
/	Page 141. Produces a slash / in text and in math mode. In the latter case it is treated as an ordinary symbol for purposes of spacing, but it can grow with `\big` as if it were a delimiter: pages 146–147. No break is allowed after a / not followed by space: compare `\slash` /. Also useful with abbreviations: page 166.
`\/`	Page 37. Introduces an italic correction, a bit of spacing to compensate for the slant of the previous letter. Useful after italicized and slanted words in the middle of upright text.
:	Produces a colon in text and in math mode (page 134). In the latter case it's considered a relation and automatically gets a thick space before and after: $f : X \rightarrow Y$. But `$x:=y$` gives $x := y$, which is the right thing: the spacing is placed before and after the relational symbols : and =, but not between them. See also `\colon`. For use in writing French, see page 174. See also `\spacefactor`.
;	Produces a semicolon in text and in math mode. In the latter case it's automatically followed by a thin space: $H(X, Y; X', Y')$. See also `\spacefactor`.
`\;`	Math mode only. Leaves a thick space⎸equal to 5 math units, stretchable to 10. This space is automatically placed around relations (page 134), but its explicit use is rare.
<	In math mode, produces the less-than sign <, a relation (page 134); it also serves as an abbreviation for `\langle` after `\bigl`, `\left`, etc. (page 146). Outside math mode it should only be used with typewriter fonts, otherwise it appears as a strange character ¡. Also used between two numbers or dimensions in the conditional tests `\ifnum` and `\ifdim` (pages 181–182), in which case it doesn't print.
`\<`	Not part of plain T_EX. Defined and used on page 128 to typeset the following word in bold.
∠, ⟨	See `\angle`, `\langle`.
=	In math mode, produces the less-than sign =, a relation (page 134); see also `\Relbar` =. Also used between two numbers or dimensions in the conditional tests `\ifnum` and `\ifdim` (pages 181–182) and, optionally, in all assignments: see assignments, space tokens. In such cases it doesn't print.
`\=`	Page 20. Places a bar, or macron accent, above the following character: ā. Works only in text mode; for math mode, see `\bar`. For a bar over several characters, see `\overline`.
>	In math mode, produces the greater-than sign >, a relation (page 134); it also serves as an abbreviation for `\rangle` after `\bigr`, `\right`, etc. (page 146). Outside math mode it should only be used with typewriter fonts, otherwise it appears as a strange character ¿. Also used between two numbers or dimensions in the conditional tests `\ifnum` and `\ifdim` (pages 181–182), in which case it doesn't print.
`\>`	Math mode only. Leaves a medium space⎸equal to 4 math units, stretchable to 6 and shrinkable to 0. This space is automatically placed around binary operators (page 133), but its explicit use is rare.

)   See `\rangle`.

?   Produces a question mark in text and in math mode. See also `\spacefactor`.

?'   Page 19. Produces the Spanish ¿.

@   Produces an at-sign @ in text and in math mode. Its category can be changed so it can be made part of the name of "protected" control sequences: see page 177 and an example under `\afterassignment`.

[Produces a left bracket in text and in math mode. As a math delimiter, it can grow arbitrarily large with `\left` (pages 146–147). See also `\count`.

\[Not part of plain TEX. Defined and used on page 128 to typeset the following word in italics.

\\   Pages 164, 172. The escape character: introduces a control sequence, a word that isn't printed but interpreted as a command. To print a \ in math mode, use `\backslash` or `\setminus`. In text you must either switch to a typewriter font and say `{\tt\char'\\}`, which gives \, or go temporarily into math mode.

]   Produces a right bracket in text and in math mode. As a math delimiter, it can grow arbitrarily large with `\right` (pages 146–147). See also `\spacefactor`.

^   Pages 17–18, 136, 172. In math mode, introduces a superscript: see superscripts, superscript character. See also page 149 and `^^`.

\^   Page 20. Places a circumflex accent over the following character: ê. Works in text mode only; for math mode, see `\hat`. For î you must type `\^\i` rather than `\^i`. The discussion about `"` on page 174 is relevant if your text has many accents.

^^   Gives a way to represent unprintable ASCII characters in a source file: `^^` followed by a character whose ASCII code is n represents the character whose ASCII code is $n + 64$, if $n < 64$, or $n - 64$, if $64 \leq n < 128$. For example, a carriage return (CR) can be written `^^M` (page 175), because its code is 13 and M's code is 77. This substitution works even inside a control sequence name.

You can also refer to any character with code between 0 and 255 by saying `^^`xy, where xy stands for the character code in hexadecimal (letters must be lowercase). This convention overrides the one described in the previous paragraph, that is, `^^a0` represents the character with code 11×16, not the character `^^a` followed by 0. This feature is not available in versions of TEX prior to 3.0.

The following non-printable characters are discussed in this book: `^^@`, or NULL; `^^A`, or SUB; `^^I`, or TAB; `^^J`, or SUP; `^^M`, or CR; and `^^?`, or DEL.

_   Pages 17–18, 136, 172. In math mode, introduces a subscript: see superscripts. See also page 149.

_   Produces an underscore _, a character mostly used by computer scientists and programmers.

         `\it very\_long\_identifier\_name` .... *very_long_identifier_name*

'   Page 18. Produces left single quotes ' in text; see also quotes.

When TEX is expecting to read an integer (page 181), you can use ' followed by a character token (page 190) or a control sequence token whose name has only one character. The resulting integer value is the ASCII code of the character. Thus `\catcode'\$` is the same as `\catcode 36`, because 36 is the ASCII code of $; that's why the construction of page 172 works. See also integers.

\`  Page 20. Places a grave accent over the following character: è. Works in text mode only; for math mode, see \grave. For ì you must type \`\i rather than \`i. The discussion about " on page 174 is relevant if your text has many accents.

` ` Pages 18–19. Produces left double quotes " in text; see also quotes.

`\  See `.

{ Pages 17, 22, 172. The most common way of starting a group: changes made inside the group are canceled when the matching } is found. Braces also follow such commands as \vbox and \halign; this use also causes grouping. Yet another use of braces is to delimit the definition of a macro (page 165), or an "undelimited" macro argument (pages 167–168, 170), or a token list (page 178).

Unmatched braces are a common input error (pages 26, 171): hence the idea of coding from the outside, braces first (pages 105, 152, etc.) In macro definitions, on the other hand, unmatched braces can be very useful, but they must sneak in in disguise: see \bgroup.

To print braces, see braces, \{.

\{ Page 135. Produces a left brace in math mode. As a math delimiter, it can grow arbitrarily large with \left (pages 146–147).

| Page 134. Produces a vertical bar, also obtained with \vert. As a math delimiter, it can grow arbitrarily large with \left and \right. The same symbol, with spacing on both sides, represents a relation: in that case it should be coded \mid (normal size), \bigm|, and so on. See pages 146–147.

In this book, | is used to introduce verbatim mode (page 174).

\| Page 134. Produces a double vertical bar ‖, also obtained with \Vert. As a math delimiter, it can grow arbitrarily large with \left and \right. The same symbol, with spacing on both sides, represents a relation: in that case it should be coded \parallel (normal size), \bigm\|, and so on. See pages 146–147.

} Pages 17, 22, 172. Ends a group started by {; see that entry for more information.

\} Page 135. Produces a right brace in math mode. As a math delimiter, it can grow arbitrarily large with \right (pages 146–147).

~ Pages 18, 21, 39, 45, 76, 173. Creates a tie, a space at which TEX will not break lines and which is the same regardless of what the preceding character (see \spacefactor). The ~ replaces the space in the input; thus you should say p.~314, rather than p.~ 314.

\~ Page 20. Places a tilde over the following character: ẽ. Works in text mode only; for math mode, see \tilde. For ĩ you must type \~\i rather than \~i. The discussion about " on page 174 is relevant if your text has many accents.

\aa, \AA Page 19–20. Produce the Scandinavian letter å, Å.

abbreviations in mathematics, page 143; in a preamble, page 116; in typing, page 166; see also macros.

\above Page 142. Math mode only. Like \over, but makes a fraction bar of specified thickness:

$$x+{y+z\above 1pt v+w}$$ .... $x + \dfrac{y+z}{v+w}$

\abovedisplayshortskip, \abovedisplayskip
  Page 131–132. Two of TEX's glue variables: they control the amount of vertical spacing between a displayed math formula and a preceding line of text in the same paragraph. The

first value is used if the end of the line is at least two quads to the left of the beginning of the formula. Plain TeX sets

```
\abovedisplayshortskip=0pt plus 3pt
\abovedisplayskip=12pt plus 3pt minus 9pt
```

See also `\eightpoint` .

\abovewithdelims

Page 142. Math mode only. Like `\over` , but makes a fraction bar of specified thickness, and encloses it between specified delimiters:

$$\texttt{\$\$x+\{y+z\textbackslash abovewithdelims[] 1pt v+w\}\$\$} \dots\dots\dots\dots\dots x + \left[\frac{y+z}{v+w} \right]$$

Watch out! The construction `\abovewithdelims[.1pt` doesn't mean the bar thickness is .1 pt: the . is a dummy closing delimiter (cf. page 147).

accents   in text, pages 7, 20; in mathematics, page 137.

\active   Pages 111, 173. A name for the number 13, the category code of an active character. For example, plain TeX says `\catcode`\~=\active` prior to defining ~ as a tie.

active characters

Pages 111, 173–176. A character of category 13 is treated by TeX as a macro; its meaning must be defined with `\def` before the character can be used. Once defined, the meaning is available whenever the character is active, no matter how many times the character has been deactivated and reactivated in between. See also `\mathcode` .

\acute   Page 137. Math mode only. Places an acute accent over the following character: $\acute{a}$. Its text counterpart is `\'` .

\adots   Page 160. Not part of plain TeX. Math mode only. Makes three ascending dots $\cdot\cdot\cdot$.

Addison-Wesley

Pages 1–2.

addition   Page 183.

address formatting

Pages 54, 71.

\advance   Page 183. Adds to a register of type integer, dimension, glue or math glue another object of the same type. Most commonly used to automatically increment counters for footnotes, page numbers, and such like: see page 62 and the next entry.

\advancepageno

This macro advances the value of `\pageno` variable when it comes time to move on to a new page:

```
\def\advancepageno{\ifnum\pageno<0\global\advance\pageno by -1
    \else\global\advance\pageno by 1\fi}
```

The point of this is that, by convention, a negative `\pageno` is printed (by the `\folio` macro) as a roman numeral. To increase the page number, then, `\pageno` should be added 1 if positive, and added −1 if negative. The `\global` is necessary because `\advancepageno` is ordinarily called inside a group, during the output routine.

\ae, \AE   Page 19. Produce the Scandinavian and Latin ligature æ, Æ.

Æsop   Page 19.

aesthetics   Pages 1, 3–4, 7; see also fine-tuning

\afterassignment

> Stores the next token until an assignment is performed, then puts it back into the input stream. Useful in writing macros whose "arguments" don't have to be placed in braces. Consider this bit of code from plain TEX:
>
> ```
> \def\openup{\afterassignment\@penup\dimen@=}
> \def\@penup{\advance\lineskip\dimen@
>   \advance\baselineskip\dimen@
>   \advance\lineskiplimit\dimen@}
> ```
>
> When TEX sees \openup, it assigns to the register \dimen@ the dimension that follows. Immediately after that, it sees \@penup, whose expansion causes \dimen@ to be added to \lineskip, \baselineskip and \lineskiplimit. See \romannumeral for another example.

\aftergroup Stores the next token until the current group ends, then puts it back into the input stream. Useful in writing macros whose "arguments" are not read ahead of time. Suppose, for example, that you want \toto{...} to expand to \pre{...}\post, but you want the stuff in braces to be read only *after* \pre, perhaps because \pre changes some category codes. If you say \def\toto#1{\pre#1\post}, TEX will read and store the argument before \pre has a chance to act. Here's a solution:

> ```
> \def\toto{\pre\bgroup\aftergroup\post\let\dummy=}
> ```

> The trick is to get rid of the left brace after \toto, by assigning it to a dummy control sequence. The right brace at the end of the argument balances with the \bgroup, and its occurrence causes TEX to process \post. Naturally, \pre and \post can contain anything, including additional "braces" \bgroup and \egroup.

> This solution assumes that the argument to \toto will always come in braces. It is possible to relax that assumption by looking at the character following \toto, using \ifnextchar, for example. But it complicates things.

> For an application, see \myfootnote.

\aleph   Page 132. Math mode only. Produces the Hebrew letter ℵ.

aligning   boxes, pages 90, 99, 101; digits, page 111; formulas, pages 154, 187; radicals, page 144; subscripts and superscripts, page 136.

alignments   Pages 102–129; exploded view of, pages 102, 114; and group nesting, pages 104, 157; in macros, page 169; in math, pages 147, 154–155, 157–163, 186–189; opening up, pages 112–113, 152, 154–155, 157, 186; spanning columns of, page 128; spillover entries, pages 108, 123.

alignment separator
>   Page 172. Any character of category 4, but generally &.

allocating   a box, page 84; a register, page 179.

\allowbreak Page 76. An abbreviation for \penalty 0. It gives TEX permission to break a line or page at a point where it would not consider it otherwise. Useful in long math formulas:

> ```
> $(x_0,\dots,x_{i-1},\hat x_i,\allowbreak x_{i+1},\dots,x_n)$
> ```

\alpha   Pages 132, 185. Math mode only. Produces the Greek letter α.

alphabetic constants
>   See `.

assignment   Page 179; see also = . Examples of assignments are given on pages 29, 64, 84, 166, 173, 178.

\ast   Page 153. Math mode only. Produces the binary operator ∗. The character ∗ has the same effect.

asterisk   See ∗.

\asymp   Page 134. Math mode only. Produces the relation ≍.

at   Page 30. Keyword used in registering a font, if the font is to be used at other than its design size: \font\bigten=cmr10 at 12pt . See examples on pages 92, 96.

at sign   See @ .

\atop   Pages 142, 147. Math mode only. Stacks two formulas atop one another, as if to form a fraction, but without the horizontal bar. Here's a non-obvious use: the matrix $\left(\begin{smallmatrix} a & c \\ b & d \end{smallmatrix}\right)$, which fits comfortably in a paragraph, isn't easy to generate with \matrix . Use instead $\bigl({a \atop b}{c\atop d}\bigr)$.

\atopwithdelims

Page 142. Math mode only. Variant of \atop that surrounds the stacked formulas with delimiters:

$$q+{x+u \atopwithdelims<> x+vw}$$ . . . . $q + \left\langle \begin{matrix} x+u \\ x+vw \end{matrix} \right\rangle$

Used rarely, because there are abbreviations \choose , \brace and \brack for the most common cases.

author   See \runningauthor .

avatars of T$_E$X   Page 5.

axis   Pages 100–101, 144, 148. An imaginary horizontal line used as reference for the placement of a \vcenter box and of many mathematical symbols, like $-$, $+$, $\cdot$, and fraction bars. The axis height is stored in the register \fontdimen22\textfont2 . To lower something that's centered about the axis so it's centered about the baseline, you must say

$$\lower\fontdimen22\textfont2\hbox\{...\}$$

The same construct with \raise achieves the opposite effect.

\b   Page 20. Places a bar under the following character: e̠. Works in text mode only; for math mode, or for more than one character in text, see \underline .

backslash   See \ , \backslash , \setminus .

\backslash   Page 133. Math mode only. Produces a backslash \. The same symbol is obtained with \setminus , but the spacing is different:

$H\backslash G/K$, $X\setminus Y$ . . . . $H\backslash G/K,\ X\setminus Y$

backspacing   Pages 39, 44, 48, 61, 123, 159.

badness   Page 41; see also overfull, underfull, \hbadness , \vbadness , \pretolerance , and \tolerance .

Baillie, Kate   Page 60.

balance of power
     Page 77.

balancing columns
     Pages 98.

bars   Pages 20, 145–146; see also rules and the next entry.

\bar   Page 137, 145. Math mode only. Places a small bar over the following character: $\bar{a}$. Its text counterpart is \= . For a longer bar, or a bar over several characters, see \overline .

baseline   Pages 78–80, 86–87, 97–98, 100, 143.

\baselineskip
                Page 93. One of TEX's glue variables: it controls the normal distance between baselines of consecutive boxes or lines of text that are being stacked vertically. The value used to separate the lines of a paragraph is the value at the end of the paragraph, so if you change \baselineskip inside a group you must say \par before closing the group, or the last paragraph won't be affected (cf. page 54).

                Plain TEX sets \baselineskip=12pt , for use with 10-point fonts. You can modify this value directly or indirectly, using \openup . The \normalbaselines macro resets the default value; some macros like \matrix include a call to it (pages 159–160, 186). See also pages 112–113 for its effect on tables.

Basho, Matsuo
                Page 53.

Basic   Page 65.

\batchmode   Causes TEX to run without stopping for errors, no matter how serious, and not to print anything on your screen. Typing q in response to an error has the same effect (page 15). See also error checking.

\bb   Pages 34–35. Not part of plain TEX. Some mathematicians prefer using "blackboard bold" letters $\mathbb{Z}$ and $\mathbb{R}$ instead of **Z** and **R**, even when they're not at the blackboard. If you're one of them, try

```
\font\tenbb=msbm10   \textfont\bbfam=\tenbb
\font\sevenbb=msbm7  \scriptfont\bbfam=\sevenbb
\font\fivebb=msbm5   \scriptscriptfont\bbfam=\fivebb
\newfam\bbfam        \def\bb{\fam\bbfam}
```

                After that, $\bb R$ will give $\mathbb{R}$, and so on. This assumes your system has the appropriate fonts msbm10 , etc., which are not part of the Computer Modern family; if not, you can get them from the American Mathematical Society. (The fonts come with macro files amssym.def and amssym.tex that define a command \Bbb that is essentially the same as \bb , so if you \input this file you don't have to worry about the definition above.)

\bbuildrel   Pages 149, 170. Not part of plain TEX. Like \buildrel , but lets you add labels both above and below an arrow or other relation.

beauty   See aesthetics.

beads   Pages 24, 78–79.

\begin   Pages 24. Not part of plain TEX. This LATEX macro starts an *environment*, which must be ended by a corresponding \end command. The \midinsert...\endinsert construction of plain TEX might be written in LATEX \begin{insert}...\end{insert} .

begin/end-math
                Page 172. Any character of category 3, but generally $. Any begin/end-math character can terminate a group started by another.

begin-group character
                Page 172. Any character of category 1, but generally { . Any end-group character can match

any begin-group character. Therefore begin- and end-group characters must be balanced inside a macro definition; to define a macro that starts a group but doesn't end it, you must resort to \bgroup .

\begingroup Pages 23–24. Marks the beginning of a special-purpose group, which must be ended by \endgroup . Mostly used in the definition of macros in certain packages that do error checking: otherwise { and \bgroup are sufficient.

beginning a paragraph
  See mode changes.

\beginsection

A plain TEX command to start a new section in a document. The section title is delimited by a \par or empty line, and printed in bold: \beginsection Conclusion \par The optimist ... gives

Conclusion

The optimist thinks we live in the best of possible worlds. The pessimist agrees.

This macro is \outer , that is, it is not allowed to appear in macro definitions and in certain other situations. To get around this, see \outer .

See also \section .

\belowdisplayshortskip, \belowdisplayskip

Page 131–132. Two of TEX's glue variables: they control the amount of vertical spacing between a displayed math formula and a following line of text in the same paragraph. The first value is used if the end of the line preceding the formula is at least two quads to the left of the beginning of the formula.

  \belowdisplayshortskip=7pt plus 3pt minus 3pt
  \belowdisplayskip=12pt plus 3pt minus 9pt

See also \eightpoint .

Benares   Page 155.

Berber   Page 21.

Bergstrøm   Page 19.

\beta   Page 132. Math mode only. Produces the Greek letter β.

\bf   Pages 28, 34–37, 135. Switches to a boldface font, in text or in math mode. Normally should be used inside a group, so its effect goes away when the group ends.

  In plain TEX \bf always switches to the text font \tenbf , and to the math family \bffam . To set things up so that \bf switches to a boldface font in the current size, see \eightpoint and \tenpoint .

\bffam   Page 34–36. A name for the boldface font family to be used in math mode. To select that family, say \fam\bffam . The \bf command does this.

\bgroup   Pages 23–24. Another name for the left brace { . Its use is necessary for unmatched braces inside a macro definition or token list: see an example on page 167.

bibliographies Pages 21, 37, 136, 169.

bicycle mechanics
  Page 99.

Bilbo Baggins Page 11.

\big, \Big Pages 146–148. Math mode only. \big makes a following delimiter slightly bigger, but not so big that it will disturb the spacing between lines in a paragraph; while \Big makes it half again as big as the corresponding \big delimiter. Here are all of TEX's delimiters, in regular size, \big size and \Big size:

These commands are commonly used only with / and \backslash. The variants \bigl, \bigm, \bigr and corresponding \Big ones are to be preferred, because they put the right amount of spacing around the delimiter.

\bigbreak Page 77. Causes a conditional vertical skip by \bigskipamount, and marks the place as a very good one for a page break. If the \bigbreak was preceded by another skip, the lesser of the two is canceled; in particular, two \bigbreak s have the same effect as one.

\bigcap Page 133. Math mode only. Produces the large operator $\bigcap$, $\bigcap$. Compare \cap $\cap$.

\bigcirc Page 133. Math mode only. Produces the binary operator $\bigcirc$. Compare \circ $\circ$.

\bigcup Page 133. Math mode only. Produces the large operator $\bigcup$, $\bigcup$. Compare \cup $\cup$.

\bigg, \Bigg

Pages 147–148. Math mode only. \bigg makes a following delimiter twice as big as the corresponding \big delimiter, while \Bigg makes it two and a half times as big:

These commands are commonly used only with / and \backslash. It's preferable to use the variants \biggl, \biggm, \biggr and corresponding \Bigg ones, because they put the right amount of spacing around the delimiter.

\biggl, \Biggl

Pages 146–147. Math mode only. Like \bigg and \Bigg, but used before an opening delimiter.

\biggm, \Biggm

Pages 146–147. Math mode only. Like \bigg and \Bigg, but used before a relation (generally a vertical bar or arrow).

\biggr, \Biggr

Pages 146–147. Math mode only. Like \bigg and \Bigg, but used before a closing delimiter.

\bigl, \Bigl

Pages 146–147. Math mode only. Like \big and \Big, but used before an opening delimiter.

\bigm, \Bigm

Pages 146–147. Math mode only. Like \big and \Big, but used before a relation (generally a vertical bar or arrow).

\bigodot   Page 133. Math mode only. Produces the large operator $\bigodot$, $\odot$. Compare \odot $\odot$.

\bigoplus   Page 133. Math mode only. Produces the large operator $\bigoplus$, $\bigoplus$. Cf. \oplus $\oplus$.

big point   See bp .

\bigr, \Bigr

> Pages 146—147. Math mode only. Like \big and \Big , but used before a closing delimiter.

\bigskip   Page 40. Causes a vertical skip by \bigskipamount . See also \bigbreak .

\bigskipamount

> One of TEX's glue variables: it controls the amount of a \bigskip . Plain TEX sets it to 12pt plus 4pt minus 4pt . See also \eightpoint .

\bigsqcup   Page 133. Math mode only. Produces the large operator $\bigsqcup$, $\bigsqcup$. Compare \sqcup $\sqcup$.

\bigtriangledown

> Page 133. Math mode only. Produces the binary operator $\bigtriangledown$. (There is no corresponding \triangledown .)

\bigtriangleup

> Page 133. Math mode only. Produces the binary operator $\triangle$, which is the same symbol as \triangle , but with spacing on both sides. Compare also \triangleleft $\triangleleft$ and \triangleright $\triangleright$ (there is no \triangleup .)

\biguplus   Page 133. Math mode only. Produces the large operator $\biguplus$, $\biguplus$. Compare \uplus $\uplus$.

\bigvee   Page 133. Math mode only. Produces the large operator $\bigvee$, $\bigvee$. Compare \vee $\vee$.

\bigwedge   Page 133. Math mode only. Produces the large operator $\bigwedge$, $\bigwedge$. Compare \wedge $\wedge$.

binary operator

> Page 131, 133, 153. A symbol or subformula of class 2. A binary operator preceded and followed by letters or subformulas is separated from them by a medium space (given by \medmuskip). But if nothing precedes the operator in the same subformula, or if nothing follows it, no spacing is placed either before or after it: the operator is assumed to be unary. Thus $(-x)$ gives $(-x)$, while ${(\{\}-x)}$ gives $(- x)$. As with any symbol, you can deprive a binary operator of spacing altogether by placing braces around it. Also, the spacing is never put in in script and scriptscript styles. See also \mathbin .

binomial coefficients

> See \choose .

birds, extant and extinct

> Page 109.

bitmap fonts Page 31.

blackboard bold

> See \bb .

black box   See \overfullrule .

blank   box, pages 46, 89; line in input, pages 38, 52, 54, 171; page, page 69; space, see SP, spacing.

\bmod   Math mode only. Produces the abbreviation mod, considered as a binary operator. To be distinguished from \pmod , which prints a "modulo condition" in parenthesis:

> $y=x\bmod a$ .... $y = x \bmod a$
>
> $x\equiv y\pmod n$ .... $x \equiv y \pmod n$

boldface   Pages 27–29.

book.mac   Pages 50, 174. File containing macro definitions and style settings for this book.

boolean variables
                  See \if , \newif .

\bordermatrix
                  Math mode only. Creates a matrix bordered with labels. An $n \times n$ border matrix is input
                  like a normal $(n + 1) \times (n + 1)$ matrix with the upper-left corner entry empty. TEX places
                  parentheses at the right places.

$$A=\bordermatrix\{$$
$$\& \ p \ \& \ q \quad \backslash cr$$
$$p \ \& \ I\_p \ \& \ 0 \quad \backslash cr$$
$$q \ \& \ 0 \quad \& \ J\_q \ \backslash cr \ \}$$

$$A = \begin{matrix} & p & q \\ p & \\ q & \end{matrix} \begin{pmatrix} I_p & O \\ O & J_q \end{pmatrix}$$

Boston Computer Museum
                  Page 1.

\bot   Page 132. Math mode only. Produces $\perp$, the same character as \perp , but without spacing
                  around it.

\botmark   The mark text most recently encountered on a page that TEX has just completed. See \mark .

Bourbaki, Salocin
                  Page 71.

\bowtie   Page 134. Produces the relation $\bowtie$, made with \joinrel from \triangleright and
                  \triangleleft .

boxes   Page 78; arrangements of, pages 89–90, 97–98, 101; baseline of, pages 86–87; blank,
                  page 89; centered, see \vcenter ; character, pages 78, 80, 86; corner, page 92; dimensions
                  of, pages 79–80, 88–89, 91; explicit, page 83; framed, see \boxit ; and glue, pages 81,
                  93; and groups, pages 24, 83; horizontal, page 80; lowering, page 87; moving right and left,
                  page 88; placing side by side, page 90; raising, page 87; register, pages 84, 178–179; and
                  rules, page 81; splitting, page 98; storing, pages 84, 178–179; unboxing, page 85; using,
                  page 85; vertical, page 79; void, page 185.

\box   Pages 85, 178–179, 181. Uses a previously stored box, and erases it: after

                  $$\setbox2=\hbox\{...\}\box2$$

                  box 2 is void. This erasure is not an assignment: it works across groups, up to the level
                  where the box was assigned to. In other words, replacing \box2 by {\box2} in the line
                  above wouldn't preserve the contents, even outside the group.
                  The \box command must be followed by an integer between 0 and 255, either explicit or
                  symbolic (i.e., an integer register or variable). Except for boxes 0 through 9, which can be
                  used for temporary storage, all other boxes should be allocated using \newbox : pages 84,
                  179.

\boxit   Page 97. Not part of plain TEX. Puts its argument into a box and draws a frame around it,
                  leaving a specified amount of white space: \boxit{2pt}{toto} gives $\boxed{\text{toto}}$.

bp   A keyword for big point, one of TEX's units. A big point is just a tiny bit bigger than a
                  regular point pt : by definition, one inch equals 72 big points.

\brace   Page 142. Math mode only. An abbreviation for \atopwithdelims\{\} :

                  $$p+\{n \ \brace k\}\$$ .... $p + \left\{ \begin{matrix} n \\ k \end{matrix} \right\}$

braces | in input, see { , groups, \bgroup ; horizontal, pages 46, 145; vertical, pages 135, 146, 155.

\brack | Page 142. Math mode only. An abbreviation for \atopwithdelims[] :

$$\$\$p+\{n \ \backslash brack \ k\}\$\$ \dots\dots\dots\dots\dots\dots\dots\dots\dots\dots\dots p + \begin{bmatrix} n \\ k \end{bmatrix}$$

brackets | See [.

\break | Page 76. An abbreviation for \penalty -10000 . It forces T_EX to start a new line, if in horizontal mode, or a new page, if in vertical mode. A line break does not start a new paragraph. \break doesn't add any glue, so generally it makes the line or page underfull unless you precede it with \hfill or \vfill. Also, xyz \break introduces a space after xyz. For examples of use, see pages 46, 59, 70, 76, 95.

breaking | lines, pages 41, 44, 53, 70, 75–77; pages, pages 41, 44, 75–77; tables, page 105; formulas, see page 153, * , \allowbreak . See also \break , hyphenation, discretionary, fine-tuning.

\breve | Page 137. Math mode only. Places a breve accent over the following character: ă. Its text counterpart is \u .

Brezé, Pierre de
| Page 60.

Buddha | Page 155.

\buildrel | Pages 148, 170. Places a symbol or group above another, as if superscripted: $\buildrel \rm def \over{=}$ gives $\stackrel{\text{def}}{=}$. The material between \buildrel and \over forms a group by itself, but a group after the \over must be explicitly marked. See also \mathop , \bbuildrel .

bugs | Page 5.

\bullet | Pages 50, 133. Math mode only. Produces the binary operator •, also commonly used with the \item and \meti macros.

by | Page 183. Keyword used with the arithmetic operations \advance , \multiply and \divide : \advance\pageno by 1 . In each case, by can be omitted.

\bye | Pages 16, 76. The recommended way to end a T_EX run: see examples on pages 11–12, 25–26, 75.

\c | Pages 20, 21. Places a cedilla under the following letter: Fran\c cois yields François. Not for use in math mode.

C version of T_EX
| Page 6.

caddish behavior
| Page 18.

\cal | Pages 130, 136, 149. Math mode only. Switches to a font that has calligraphic capitals: $\cal ABC$ gives $\mathcal{ABC}$. If you don't put it inside a group the whole formula is typeset in this font, with ludicrous results:

$$\$\backslash cal \ A*b+c*T+x/y-z\$ \dots\dots\dots\dots\dots\dots\dots\dots\dots\dots\dots \mathcal{A} * \lfloor + \rfloor * \mathcal{T} + \S/\dagger - \ddagger$$

\cap | Page 133. Math mode only. Produces the binary operator ∩. Compare \bigcap ⋂.

capitals | See \uppercase . For caps and small caps, see pages 29, 169.

captions | See legends.

caret | See ^ , \^ .

carriage return
> See CR, \cr .

carrot and stick
> See penalties.

\cases   Page 147, 159, 169, 186. Math mode only. For the enumeration of two or more cases in formulas:

```
$\varphi(x) = \cases{
0         & if $x\leq 0$,\cr
e^{-1/x} & otherwise.            \cr}
```

$$\varphi(x) = \begin{cases} 0 & \text{if } x \leq 0, \\ e^{-1/x} & \text{otherwise.} \end{cases}$$

On each row, the material before the **&** is typeset in math mode, and the material after in horizontal mode. To separate the rows, use \noalign{\smallskip} immediately after each \cr , because \openup has no effect.

\catcode   Pages 172–173, 175–177. Followed by the ASCII code of a character or a construction of the form `\X, functions as an integer register that holds the character's category (see pages 178–179 for registers). For example, to change the category of @ to "letter" you say \catcode`\@=11 ; to save the current category in a register, \oldcode=\catcode`\@ ; to print it on the screen, \showthe\catcode`\@ .

category   Pages 172–173. Number between 0 and 15 indicating what function a character currently serves. See also \catcode , \active , active characters.

cc   A keyword for cicero, the pica's counterpart in most European countries. A cicero is 7% bigger than a pica.

\cdot   Page 133. Math mode only. Produces the binary operator $\cdot$, which is vertically centered:

$x\cdot y=x_1y_1+\cdots+x_ny_n$ . . . . $x \cdot y = x_1 y_1 + \cdots + x_n y_n$

\cdotp   Same as \cdot , but treated as punctuation. Sometimes useful after a fraction in display math mode:

$x={1\over 2}+{1\over 3}.$ . . . . $x = \dfrac{1}{2} + \dfrac{1}{3}.$

$x={1\over 2}+{1\over 3}\cdotp$ . . . . $x = \dfrac{1}{2} + \dfrac{1}{3} \cdot$

To raise other punctuation marks, see axis.

\cdots   Pages 145, 160. Math mode only. Produces three dots at the right height to be placed between operators such as $+, -, \times, =, <, >, \subset$ and $\supset$. See an example under \cdot .

cedilla   See \c .

centering   alignment entries, page 104; numerator or denominator, page 43; tables, pages 105, 119, 123; text line by line, page 57; vertically, page 101. See also the next entry and $$.

\centerline Page 68. Centers what follows in braces: \centerline{\bf Chapter 2} gives

Chapter 2

\centerline works by creating a box the same width as the page, so it must be used in vertical mode. If you use it inside a paragraph, you'll get an overfull box. \centerline uses springs \hss that can stretch or shrink indefinitely, so the centered material can spill over the margins without TEX complaining about an overfull box. See examples on pages 12, 15, 16, 49–50, 69–70, 89–90, 119.

centimeter   See cm .

Chandler, Raymond
> Page 66.

\chapter Page 51. Not part of plain T_EX. Macro used in this book to start a new chapter; compare \section and \subsection. It takes two arguments, the chapter number and the title. The definition on page 51 was a fib; the real definition is the following:

```
\def\chapter#1#2{\vfil\eject
    \message{Chapter #1. #2}
    \chapno=#1 \sectno=0 % for automatic section numbering
    \chaptitle={#2}       % used in \oddpagehead
    \notenumber=0         % for automatic footnote numbering
    \rightline{\chapnumfont #1}\medskip
    \rightline{\chaptitlefont #2}\vskip 14pc
    \hrule height 0pt     % to prevent \section from
                          % deleting the \vskip
    \pagetitletrue}       % no header or footer
```

\char Page 36. Followed by an integer $0 \leq n < 256$, prints the character in position n of the current font. The integer can be expressed in any notation, or come from a register, etc. (see page 181). In text fonts the position of ASCII characters normally occurring in text generally corresponds to their ASCII codes. For example, the ASCII code of 'a' is 97, so \char97 or \char'141 or \char'61 or \char'\a all print a if the current font is cmr10, plain T_EX's basic font.

The contents of other slots in the fonts is unpredictable: \char 15 or \char'017 or \char"F or \char'\^^0 prints ffl if the current font is cmr10, but ¿ if the font is cmtt10. One common case is \char'\\, which prints a backlash if the font has one, as is the case with typewriter fonts, but prints " with cmr10.

A character token of category letter (11) or ordinary (12) similarly causes T_EX to print whatever is in the position of the current font that is given by the character's ASCII code. Thus typing \char'\$X$ gives the same result as {\catcode'\$X$=12 X}. T_EX's internal operations \the, \string, \number and the like create tokens of category 12, so \string X will again produce the same result (unless X is an escape character, in which case T_EX will read in a control sequence token and apply \string to that).

characters active, pages 111, 173–176; as boxes, pages 78, 80, 86; category of, pages 173; dimensions of, pages 9, 80; in vertical mode, pages 25, 52, 82, 88, 109, 120. See also \char.

\chardef Page 36. Makes the following control sequence an abbreviation for a \char construction: after \chardef\&='\&, the control sequence \& gives the same result as \char'\&. This is how \& is defined in plain T_EX. A control sequence defined with \chardef is also accepted for its numerical value, whenever T_EX is expecting an integer (page 194). See also \mathchardef.

Charolais, Count of
         Page 60.

Château Margaux
         Page 58.

\check Page 137. Math mode only. Places a háček (check mark) over the following character: $\check{C}$. Its text counterpart is \v. Some mathematicians like the same character to the right of the letter, as a superscript: the best way to do this is to define a new control sequence with \mathchardef\checkchar"7014. Then $C\checkchar$ gives $C^{\smallsmile}$.

chewing See lexical analysis.

\chi Page 132. Math mode only. Produces the Greek letter χ.

`\choose`   Page 142. Math mode only. An abbreviation for `\atopwithdelims()`:

$$\$\$p+\{n \text{ \choose } k\}\$\$ \dotfill p + \binom{n}{k}$$

`\circ`   Pages 133. Math mode only. Produces the binary operator ∘; used on page 149 to build up a new symbol. Can also be used as an exponent, to indicate degrees: `$180^{\circ}$` gives 180°.

Christmas tree ornament
    Page 144.

cicero   See `cc`.

circumflex   See `\^`, `\hat`, `\widehat`.

class of math symbol
    Pages 36, 131 and following. Controls the amount of spacing placed around the symbol.

`\cleartabs`   Page 125. Clears all tabs previously set or, if it occurs between `\+` and `\cr`, all tabs to the right of the current position.

clones   Page 165; see also `\let`.

close quotes   Page 18.

closing delimiters
    Page 131, 135; see also `\bigr`, `\right`. No space is placed between a closing delimiter and the preceding symbol, unless that symbol is a punctuation mark and the style is display or text. But spacing is placed after a closing delimiter that is followed by a large operator, a binary operator, a relation or an inner subformula, according to the rules for each of these classes. See also `\mathclose`.

club line   Page 75. The first line of a paragraph stranded on a page by itself. You can prevent club lines by setting `\clubpenalty=10000`.

`\clubsuit`   Page 132. Math mode only. Produces the character ♣.

cm   Pages 32, 39. A keyword for centimeter, one of TeX's units. By definition, an inch equals 2.54 cm.

`cmr, cmit, ...`
    See Computer Modern fonts.

code   typesetting of, page 128; see also verbatim mode.

coding hints   See discipline.

`\colon`   Math mode only. Produces a colon without any spacing before or after, unlike : , which is a relation. For example, `$f\colon g$` gives $f\colon g$. See `\mathchar` for the definition of `\colon`.

columns   Pages 102, 119, 122; common features in, pages 104, 107; glue between, page 117; of numbers, page 111; rules between, pages 114, 121, 127; width of, pages 103, 124.

`\columns`   Page 124. Used with `\settabs` to set tabs at regular intervals. For example, `\settabs 4\columns` sets tabs one-quarter of the way across the page, halfway and three-quarters of the way.

comma   See , .

commands   See control sequences, macros.

comments   Page 18; see also `%`.

\cos, \cosh, \cot, \coth

Pages 140, 143. Math mode only. Produce the abbreviations cos, cosh, cot, coth, which function as large operators without limits.

\count

Pages 178–183. Refers to one of TeX's 256 numbered registers for integers. Except for \count255 , which can be used for temporary storage, all other registers should be allocated using \newcount .

The value of \count0 is printed on the terminal when a page is output, in brackets: [1] [2] [3] ... This register is also referred to as \pageno , and ordinarily contains the page number. But \count1 to \count9 can also be used to store other levels of page numbers or subdivisions, and their contents, too, are printed on the terminal if non-zero: [1.0.4] is printed if \count0 is 1, \count1 is 0, \count2 is 4 and \count3 through \count9 are 0.

\countdef

Makes the following control sequence an abbreviation for a \count construction: after \countdef\pageno=0 , the control sequence \pageno can be used intechangeably with \count0 . In practice \countdef is almost never used, because \count registers should be allocated with \newcount : see previous entry and page 180.

CR

The basic rule on page 38 is that one carriage return character (CR) equals one space, and two or more CRs in a row start a new paragraph. The actual rule is a bit more complicated. It's here for reference, but maybe you'll never have a reason to read it. Then again, maybe you will...

From TeX's point of view your input file is divided into lines, separated by the CR character, whose ASCII code is 13. This may not be the actual character your operating system uses to separate lines, but TeX effectively starts the processing of every line by throwing out spaces SP at the right end of the line and adding a CR there.

What happens next depends on the category of CR (page 172). If its category is end-of-line, the normal case, TeX will turn it into a space, except that: (a) if the current line was empty, except perhaps for characters of category space, the CR turns into \par instead; (b) if the last thing on the line, except perhaps for characters of category space, was a control sequence made of letters, the CR is ignored.

If CR has any other category, it is treated accordingly. The most common case is to make CR active: see \obeylines .

Notice that TeX never gets to the CR if it sees a comment character while it's reading the line. Thus you can comment out active CRs (page 53).

A CR can sneak into the middle of a line in the form ^^M . If it has category 5, TeX will treat it as if it were at the end of the line, ignoring the rest of the line. This happened on page 175. TeX will treat in the same way any character of category 5.

Naturally, the CR you type to get TeX going again after an error message (page 15) or after it's shown something on the screen (page 180) is not considered part of the input. Furthermore, a line inserted from the terminal in response to an error message (page 15), starting with i or I , does not get a CR at the end.

\CR

This control sequence is read when there is a \ at the end of a line, as explained under CR. It has the same meaning as \SP .

\cr

Pages 103, 119, 122. Indicates the end of each row of a horizontal alignment obtained with \halign or \+ , or each column of a \valign . Also used in many macros that perform alignments: see pages 147, 154–158, 186–190.

Cray

Page 5.

\crcr	Page 105, 186. Turns into a \cr unless it's already placed right after a \cr or \noalign construction; in those situations it does nothing. Used in defining alignment macros.
\cup	Page 133. Math mode only. Produces the binary operator ∪. Compare \bigcup ⋃.
\csc	Page 140. Math mode only. Produces the abbreviation csc, which functions as a large operator without limits. On page 169 we redefined it to print a word in caps and small caps.
current font	Pages 39–40, 47, 66, 101, 140; see also \font , \fam .
Cyrillic fonts	Page 30. The American Mathematical Society distributes a set of Cyrillic fonts for bibliographies.
Czech accent	see \v .
\d	Page 20–21. Places a dot under the following character: Kr\d sna gives Kṛṣna. Works in text mode only.
\dag	Page 91. Produces the symbol †, in text or math mode. But see also \dagger † .
\dagger	Page 133. Math mode only. Produces the binary operator †. Unlike \dag , this responds to style changes (in subscripts, etc.), and gets spacing automatically.
dash	Page 19.
\dashv	Page 134. Math mode only. Produces the relation ⊣. Compare \vdash ⊢.
date	See \today .
\day	One of TEX's integer variables: it contains the current date (according to your computer's operating system). To print it you must preceded by \the or \number (page 180).
dd	A keyword for didot point, the point's counterpart in most European countries. A didot point is 7% bigger than an Anglo-Saxon one.
\ddag	Produces the symbol ‡, in text or math mode. But see also \ddagger ‡ .
\ddagger	Page 133. Math mode only. Produces the binary operator ‡. Unlike \ddag , this responds to style changes (in subscripts, etc.), and gets spacing automatically.
\ddot	Page 137. Math mode only. Places two dots over the following character: $\ddot{a}$. Its text counterpart is \" .
\ddots	Pages 12, 160. Math mode only. Produces three descending dots $\ddots$.
debugging	See error checking and recovery, \show and subsequent entries, \tracingcommands and subsequent entries.
decimal constants	TEX works mostly with integers, but in certain cases it understands decimal constants, as when reading a dimension. The specification of a decimal constant is simple: an optional sign, followed by optional digits, followed by . or , (continental notation), followed by optional digits. Therefore a decimal point by itself is a valid decimal constant!
decimal	comma, see the preceding entry; point, see page 132 and the preceding entry; representation, see page 181 and integers.
Decline and Fall of the Roman Empire	Page 108.
\def	Page 167–171. Next to \TeX , probably the most heavily used control sequence in this book. It defines a macro: see examples on pages 23, 35–36, 86, 105, 111, 125, 165, 173, 175. See also \edef , \gdef , \let .

\deg   Page 140. Math mode only. Produces the abbreviation deg, which functions as a large
                operator with no limits.

degrees   See \circ ∘.

DEL   Page 172. The ASCII character with code 127; plain TEX makes it of category 15 (invalid).

\delta, \Delta
                Pages 132, 151. Math mode only. Produce the Greek letter δ, Δ.

delimiter   for macro arguments, page 169; in math, pages 129, 135, 146.

demerits   Page 75.

demibold font   Page 30.

dependencies   See implementation dependencies.

depth   of boxes, see pages 78–80, \dp ; of characters, page 9; of rules, see next entry. See also
                dimensions.

depth   Page 94. An optional specification after \hrule or \vrule ; must be followed by a di-
                mension. A horizontal rule without an explicit depth will have depth zero; a vertical rule
                without an explicit depth will be as deep as the immediately enclosing box or line.

Descartes, René
                Pages 55–57.

descenders   See pages 96, 112, \underbar , \underline .

\det   Pages 140, 158. Math mode only. Produces the abbreviation det, which functions as a large
                operator, with limits in display style. To get a matrix flanked by vertical bars, write

                                    \left|\matrix{...}\right|

device   dependencies, see implementation dependencies; independence, pages 3, 9.

diagonal   arrows, pages 135, 163; dots, pages 160.

\diagram   Pages 161–163. Not part of plain TEX. Allows the creation of diagrams with arrows and
                letters.

\diamond   Page 133. Math mode only. Produces the binary operator ⋄.

\diamondsuit
                Pages 101, 132. Math mode only. Produces the symbol ◊.

dictionary of macros
                Page 165.

dieresis   See \" .

digital typography
                Page 2.

digits   Pages 17, 131, 164; aligning, page 111. See also error checking and recovery.

\dim   Pages 140. Math mode only. Produces the abbreviation dim, which functions as a large
                operator with no limits.

\dimen   Pages 178–183. Refers to one of TEX's 256 numbered registers for dimensions. Except for
                \dimen0 through \dimen9 and \dimen255 , which can be used for temporary storage,
                all other registers should be allocated using \newdimen .

\dimendef   Makes the following control sequence an abbreviation for a \dimen construction. Used
                like \countef . In practice \dimendef is almost never used, because \dimen registers
                should be allocated with \newdimen : see previous entry and page 180.

dimension   of boxes, pages 79, 88, 91; of characters, pages 9, 80; of glue, page 81; registers, pages 178–
                   183. When T_EX is expecting to read a dimension (page 181), it will accept either an "internal
                   dimension," optionally preceded by a sign, or a number followed by a unit. An internal
                   dimension is a construction of the form \dimen n, or any control sequence defined with
                   \newdimen, or a dimension variable like \parindent, or a font or box dimension (see
                   \fontdimen, \dp). A unit is a physical unit like cm (see units), optionally preceded by
                   true, or one of the font-dependent units em and ex, or an internal dimension. The number
                   preceding a unit can be any integer (see integers), or a decimal constant. Therefore .5\dp0
                   and \hangafter\baselineskip are valid dimension specifications.

                   Notice that a number by itself—even 0—is not a sufficient dimension specification. It must
                   be followed by a unit.

disappearing glue
                   Pages 44, 68.

Discours de la méthode
                   Pages 55–57.

discipline   in source file, pages 51, 104–105, 123, 131, 152, 165, 179.

discretionary   A discretionary text is a text that should be typeset differently depending on whether or not
                   it is broken across lines. Every word of more than one syllable is an example: if a word is
                   broken across lines, the first part is terminated by a hyphen. We can think of the word as
                   containing implicit discretionary hyphens, like this: won\-der\-ful, where \- typesets
                   as a hyphen or as nothing, depending on whether or not there is a break there (page 166).
                   For another example, see * and the next entry.

\discretionary
                   Implements a general discretionary text: \discretionary{ x }{ y }{ z } will print z if
                   it doesn't have to be broken, but it will print x at the end of a line and y at the beginning
                   of the next otherwise. Here x, y and z represent sequences of characters, boxes and kerns.
                   A discretionary hyphen \- (page 166) is equivalent to \discretionary{-}{}{}. Next,
                   consider the German hyphenation rule that 'ck' becomes 'kk' at a line boundary: if we set
                   \def\ck{\discretionary{k-}{k}{ck}}, dr\"u{\ck}en prints as drücken or drük-
                   ken, as the case may be.

\displayindent
                   One of T_EX's dimension variables: it controls the amount of indentation of displayed math
                   formula. See \displaywidth for more details.

\displaylimits
                   Page 143. Math mode only. After a large operator (page 133), declares that limits (or sub-
                   scripts and superscripts) should be placed above and below the operator in display style, but
                   to the right in other styles. This is the default behavior, but \displaylimits is useful
                   nonetheless to redefine an operator that has been defined with \limits or \nolimits:

                   $$\def\myint{\int\displaylimits}$$

\displaylines
                   Pages 152, 154, 157, 159, 189. Arranges several formulas in a single display, centering each
                   one.

\displaylinesno
                   Page 189. Not part of plain T_EX. Arranges several centered formulas in a single display,
                   labeling each one individually. See also \ldisplaylinesno.

display math mode

Pages 25, 105, 130; see also `\displaystyle`, `\displaywidth`.

`\displaystyle`

Pages 139. Switches to display style, the biggest of the four styles of math formulas. The change remains in effect till the end of the smallest enclosing group.

TeX starts in display style when it typesets a displayed math formula. For situations where `\displaystyle` is appropriate, see pages 139, 141, 144, 162, 187. See also `\textstyle`, `\scriptstyle`.

`\displaywidth`

Page 189. One of TeX's dimension variables: it controls the width of a displayed math formula. More precisely, such a formula is centered in a box of width `\displaywidth`, which is then placed at a distance `\displayindent` from the left margin.

TeX sets `\displaywidth` to the current line width and `\displayindent` to the current indentation just after it reads the `$$` that introduces the display. In computing line width and indentation, it takes into account the paragraph's `\parshape`, `\hangindent` and `\hangafter`, but not the values of `\rightskip` or `\leftskip`. Assume for concreteness that no hanging indentation or funny shape is in effect: the `\displaywidth` is set to `\hsize` and `\displayindent` to zero. If `\rightskip` or `\leftskip` are non-zero, this means that the display will be centered with respect to the full page, not with respect to the indented text:

$$e^{\pi i} + 1 = 0.$$

If that's not what you want, you have an opportunity to fix it, because TeX doesn't actually use the values until the end of the formula. To center a display with respect to text that has been indented with `\rightskip` and `\leftskip`, you can say

```
\advance\displaywidth by-\leftskip
\advance\displaywidth by-\rightskip
\displayindent=\leftskip
```

somewhere before the closing `$$`. Or, if you have to do this for several displays, you can say once and for all, at the top of your file:

```
\everydisplay={\advance ... \leftskip}
```

`\div`

Pages 133. Math mode only. Produces the binary operator $\div$.

`\divide`

Page 183. Divides the contents of an integer register by another integer: `\divide\pageno by2`. To divide a dimension register by two you can say `\dimen0=.5\dimen0`; this doesn't work for integers.

document description language

Page 4.

dollar sign

See `$`.

`\dot`

Page 137. Math mode only. Places a dot over the following character: $\dot{a}$. Its text counterpart is `\.`. For a dot under a character, see `\d`.

`\doteq`

Page 134. Math mode only. Produces the relation $\doteq$.

`\dotfill`

Pages 46, 98, 110, 156. Spring that leaves a trail of dots, also known as leaders.

dotless 'i' and 'j'

See `\i`, `\j`, `\mathi`, `\mathj`.

dot-matrix printer
                Page 5.

dots   Here's a table of all the dots. Take your pick, and check the individual entries.

$\cdot$ \cdot$ ⟶ ·	$\cdotp$ ⟶ ·	$\cdots$ ⟶ ···
\dots ⟶ ...	$\ldotp$ ⟶ .	$\ldots$ ⟶ ...
$\adots$ ⟶ ·˙	$\ddots$ ⟶ ··.	$\vdots$ ⟶ ⋮
\.a ⟶ å	\"a ⟶ ä	\d a ⟶ ạ
$\dot a$ ⟶ $\dot a$	$\ddot a$ ⟶ $\ddot a$	\line{\dotfill} ⟶ ....

\dots   Pages 11, 160. Produces an ellipsis, or three dots, in text... The result is better than what
                you get by typing three dots... Don't forget to say \dots\   if you want spacing after the
                dots... See also the preceding entry.

double   columns, page 8; dot, see \ddot , \" ; frame, page 97; indentation, page 57; quotes,
                page 18; subscript or superscript, page 136.

\downarrow, \Downarrow
                Pages 135, 162. Math mode only. Produce the relations ↓ and ⇓. They can be extended with
                \bigm and its bigger brothers, or with \left...\right , but in any case remain centered
                about the axis.

\downbracefill
                Pages 46, 116. A "spring" that makes braces opening down:

                \hbox to 1in{\downbracefill} .... ⏜

                Should always be put in a box by itself, because the thickness of the stroke depends on the
                height of the enclosing box (page 117); for this reason it's normally used only in alignments,
                or to match a box whose width is known:

                     \setbox2=\hbox{...}\vbox{\hbox to \wd2{\downbracefill}\box2}

                The math mode macro \overbrace does all of this for you.

\dp   Pages 88, 181, 183. The construction \dp n is like a dimension register that contains
                the depth of box n. Here n is an integer between 0 and 255, explicit or symbolic (see
                \box). You can use \dp n anywhere like a dimension register, except with \advance ,
                \multiply and \divide .

\drawbox   Pages 98. Not part of plain TₑX. Draws the outline and the baseline of a box with specified
                height, depth and width.

driver   Pages 5–7, 9, 31. Program that takes a dvi file produced by TₑX and translates the page
                descriptions in it into a language understood by a particular printer, then sends the printing
                commands to the printer.

dropped caps   Page 92.

duality   Page 119.

\dump   A TₑX run starts with the reading of a format file, like plain.tex , with definitions that
                build on TₑX's primitives. Such files can be be quite long, slowing down the startup process.
                However, TₑX is capable of encoding this startup information in a very compact file, called
                an fmt file, and of reading it back in much more quickly than it takes to read the original
                source file. The regular TₑX command, in effect, reads a file plain.fmt , automatically
                and at high speed.

If you have a format file of your own that you read in on top of plain TEX, of if you would like to replace plain TEX altogether, you may want to be able to create an `fmt` file with your definitions. To do so, you may need a special version of TEX called INITEX; consult the documentation that came with your TEX system. You start by reading in your file in the normal way: in the case of this book, we typed `initex book.mac`. After TEX has digested all the information in your file, you type `\dump` to make TEX regurgitate and end the run. In out case, this created a file `book.fmt`.

After that, we start the regular TEX by typing `&book` in response to the `**` prompt given at the beginning of each run. This makes TEX read the file `book.fmt` at high speed, and the effect is the same as if it had read the slower `book.mac` file. (Again, the exact way to read an `fmt` file depends on the implementation; in TEXtures, for example, you add an option to a menu.)

Dunhill fonts	Page 30.
`dvi file`	Page 9, 15. File where TEX places the device-independent description of the pages it has processed. The name of the file is inherited from the first input file read by TEX: for instance, `tex hobbit` will read `hobbit.tex` and create `hobbit.dvi`. If the input is interactive and no file is read on the first line of input, the output goes to `texput.dvi`.
economy	of space, page 63, 142, 153; of memory, page 85.
`\edef`	Defines a macro, expanding the replacement text at the time of the definition. Thus

$$\texttt{\edef\thispage\{\the\pageno\}}$$

defines `\thispage` to be the page number at the time of the definition, while

$$\texttt{\def\thispage\{\the\pageno\}}$$

makes `\thispage` expand to the current page number. For another example of use, see `\outer`.

editor	Page 8.
`\egroup`	Pages 23–24. Another name for the right brace `}`. Its use is necessary inside a macro definition or token list: see an example on page 167.
`\eightpoint`	Pages 35–36. Not part of plain TEX. Command to change font sizes across the board: after `\eightpoint`, an `\rm` brings in an eight-point roman, `\it` brings in eight-point italics, and so on. The listing on pages 35–36 handles all the fonts, but there is also the matter of interline spacing. For best results, add before the last line of the definition:

```
\abovedisplayskip=9pt plus 2pt minus 6pt
\belowdisplayskip=\abovedisplayskip
\abovedisplayshortskip=0pt plus 2pt
\belowdisplayshortskip=5pt plus 2pt minus 3pt
\smallskipamount=2pt plus 1pt minus 1pt
\medskipamount=4pt plus 2pt minus 2pt
\bigskipamount=9pt plus 4pt minus 4pt
\setbox\strutbox=\hbox{\vrule height 7pt depth 2pt width 0pt}%
\normalbaselineskip=9pt \normalbaselines
```

`\eject`	Pages 69–71, 76. Forces a page break. Normally preceded by `\vfill`, so the material on the current page doesn't get stretched out if the page is not full. Two `\eject`s in a row only cause one page break; to get a blank page you must say

$$\texttt{\vfill\eject\null\vfill\eject}$$

elasticity	Pages 40–41, 48, 55, 89, 93, 118, 178; see also springs, overfull, underfull.

\ell	Page 132. Math mode only. Produces the letter ℓ.
ellipsis	Page 160.
\else	Pages 183–185. Used in conditionals: \if ... \else ... \fi .
em	Pages 39, 44, 51, 140. A keyword for the em, one of TEX's units that depends on the current font.
em-dash	Page 19.
empty	box, pages 46, 89; entry (in alignment), page 106; group, pages 136, 153, 161, 168, 187.
\emptybox	Page 89, 97. Not part of plain TEX. Creates an empty (hence invisible) box of specified dimensions. For a box outline, see \drawbox .
\emptyset	Page 132. Math mode only. Produces the symbol $\emptyset$, not to be confused with the letter Ø (from \0). The font msbm10 , distributed by the American Mathematical Society, contains a version $\varnothing$ of the empty set that is closer to what one generally sees in books. For more details, see msam and msbm fonts.
Encyclopedia Britannica	Page 63.
\end	Page 16, 26. Ends a TEX run; but in some cases may not print all the insertions that are in TEX's memory. Therefore the best way to end a run is to type \bye . In LATEX \end is redefined to match \begin (page 24).
en-dash	Page 19.
end-group character	Page 172. Any character of category 2, but generally } . Any end-group character can match any begin-group character. Therefore begin- and end-group characters must be balanced inside a macro definition; to define a macro that ends a previously started group, you must resort to \egroup .
\endgroup	Pages 23–24. Marks the end of a special-purpose group started with \begingroup .
ending a paragraph	See mode changes.
\endinsert	Page 70. End of an insertion block: see \topinsert , \midinsert .
endless	See infinite.
end-of-file character	Page 16.
end-of-line character	Page 172. Any character of category 6, but generally CR. For detailed information on how TEX treats such characters, see CR.
endomorphisms	Pages 13.
engine	Page 7.
\enskip	Page 39. Leaves a space one en (half an em) wide.
\enspace	Pages 44, 59, 63, 75. Leaves a horizontal kern, or unbreakable space, one en (half an em) wide. Compare \qquad , \quad .
entries	See alignments, lists.

enumerations　Page 58.

environments　See \begin .

\epsilon　Page 132. Math mode only. Produces the Greek letter ϵ. Compare \varepsilon ε.

\eqalign　Pages 154, 159, 169. Math mode only. Arranges several formulas in a single display, aligning them vertically at the & s. The result is a \vcenter that is only as wide as the widest formula, so you can have other things in the same display as an \eqalign :

$$\left.\begin{array}{r} abc = 0 \\ a + b + c = 0 \\ ab + bc + bc = 0 \end{array}\right\} \quad \text{implies } a = b = c = 0$$

was set with $$\left.\eqalign{...}\right\}\quad\hbox{implies $a=b=c=0$}$$. See also page 147 and \cases for a similar arrangement.

Springs in the entries of an \eqalign are ineffective (page 187). The use of \noalign to insert text between the formulas of an \eqalign will destroy the centering unless the text is in an \hbox (compare page 109); in any case the text will be left-aligned with the leftmost formula, which is generally not what you want. For a better solution, see \eqalignno .

\Eqalign　Page 188. Math mode only. Not part of plain TEX. Arranges several sets of formulas in a single display, aligning each set vertically.

\eqalignno　Page 156, 159, 189. Math mode only. Arranges several formulas in a single display, aligning them vertically at the & s and labeling them on the right. The result is a box as wide as the page, so you can't have other things in the same display as an \eqalignno . On the other hand, you can use \eqalignno to maintain formulas aligned even if there is text between them:

$$\cos 2x = \cos^2 x - \sin^2 x$$

and

$$\sin 2x = 2 \sin x \cos x$$

was set with $$\eqalignno{\cos 2x&=...\cr\noalign{and}\sin 2x&=...\cr}$$. See also \leqalignno .

\eqno　Page 155. Math mode only. Labels a single displayed equation on the right; see also \leqno .

equally spaced tabs
　　　　　Page 124.

equations　See formulas.

\equiv　Page 134. Math mode only. Produces the relation $\equiv$.

error checking and recovery
　　　　　Pages 7, 10, 14–15, 24, 26, 171. When TEX stops because of an error and produces a ? followed by a message, you can respond in several ways. In each you must terminate your response with a carriage return.

(1) A CR by itself is often enough; it gets TEX going again as best it can. But the error may have immediate or delayed consequences, and you may find yourself typing CR repeatedly because of a single typo.

(2) An h causes TEX to print a help message and ask again for directions.

(3) A digit n from 1 to 9 followed by CR will cause the next n tokens in the input to be ignored. This often lets you recover more gracefully than otherwise. For example, if you forget the go into math mode and say ... `for all \epsilon>0 there is` ... , T_EX will print

```
! Missing $ inserted.
<inserted text>
                      $
<to be read again>
                          \epsilon
l. 10 ... for all \epsilon
                              >0 there is ...
?
```

It tries to insert the missing `$`. But there should be another `$` at the end of the formula, so if you respond with CR the rest of the paragraph will be set in math mode, and you'll get another error message at the end of the paragraph (cf. page 171). Instead, you should type `2`: this will delete two tokens, the inserted `$` and the offending `\epsilon`. Your formula won't come out right, but the idea is to salvage the run with a minimum of disruption, not to achieve perfection.

(4) An `i` lets you insert text: if you make a typo, say `\hobx{...}`, T_EX will tell you that the control sequence is undefined. You then have a chance to right things by typing `I\hbox`. Watch out: this correction doesn't go into the source file! It applies to this run only, and it also gets registered in the `log` file. The CR that terminates this line and any blanks that immediately precede it are not considered part of the input.

(5) An `s` tells T_EX not to stop for error messages any longer; but they still get printed on your screen.

(6) `r` is like `s`, but T_EX won't stop even for very serious errors like a missing file.

(7) `q` is like `r`, but any error messages go into the log file only, not on the screen.

(8) `x` is for exit—you throw in the towel.

As shown by the example given under (3), T_EX tries to help you figure out the source of each error by displaying a context line, broken at the point where the error occurs. If the error occurs while T_EX is expanding macros, several such broken lines are displayed, one for each level of macro expansion. Often these lines are uninformative, especially if you didn't write the macros yourself; you can control T_EX's verbosity by setting the variable `\errorcontextlines`.

\errorcontextlines

One of T_EX's token integer variables. It controls the number of context lines printed when T_EX detects an error (see the previous paragraph). If you set `\errorcontextlines=0` you only get the innermost and outermost levels of context. These are the most important anyway: the first contains the immediate cause of the error, and the last shows the ultimate cause, something that T_EX was reading from your file.

This primitive did not exist in versions of T_EX prior to 3.0.

\errorstopmode

Causes T_EX to resume its normal level of interactivity, stopping for errors. Typing CR in response to an error or interruption has the same effect (page 15).

escape character

Pages 172, 177. Any character of category 0, but normally `\`. An escape character introduces a control sequence, whose name is determined as follows: (a) If there is nothing else on the line (not even a CR), the name is empty; this can happen if you say `i\` in response to a `?` prompt. (b) If the next character is a generalized letter (has category 11), the name is made of all subsequent characters of category 11. (c) Otherwise the name is the single character after the `\`. Note: constructions like `^^M` (see `^`) are collapsed into single characters while this scanning is taking place, so `\^^M` gives a control sequence whose name has a single character `^^M` = CR.

Essai sur les mœurs et l'esprit des nations
> Page 108.

es-zet
> Pages 19, 36.

`\eta`
> Page 132. Math mode only. Produces the Greek letter η.

Euler Fraktur fonts
> Pages 30, 34. These non-Computer Modern fonts, named `eufm10`, etc., are distributed by the American Mathematical Society.

`\evenpagefoot`
> Pages 65–66, 183. Not part of plain TeX. In the `fancy` format, this variable contains the material that forms the footer of each even-numbered page. For example, if you set

$$\verb|\evenpagefoot={\hfil\tenbf\folio\hfil}|$$

you get a centered page number, in ten-point bold. It's important to set the font explicitly inside this variable, since it may be read at unpredictable moments while TeX is setting a footnote, or a caption...

`\evenpagehead`
> Pages 65–66, 71, 183. Not part of plain TeX. In the `fancy` format, this variable contains the material that forms the header of each even-numbered page. See the previous entry.

`\everycr`
> One of TeX's token list variables. It is read after every `\cr` or non-redundant `\crcr` in an alignment. Each displayed chunk of code in this book was typeset as an alignment at the "outer level" (not inside a box) with a `\noalign{\penalty500}` between rows; the idea was to make it hard, but not impossible, for TeX to break such alignments across pages. The penalties were inserted automatically by including in the macro that typesets such displays the following line:

$$\verb|\everycr{\noalign{\penalty10000}}|$$

`\everydisplay`
> One of TeX's token list variables. It is read every time TeX starts a math display, after the opening `$$`. See `\displayindent` for an example of use.

`\everypar`
> One of TeX's token list variables. It is read every time TeX starts a paragraph, that is, when it switches from vertical mode to ordinary horizontal mode. To number paragraphs automatically, you can say

```
\newcount\parcount \parcount=1
\everypar={\llap{\bf\the\parcount\quad}
\global\advance\parcount by 1 }
```

The start of each paragraph now causes TeX to print the contents of `\parcount` register on the margin, and to increment its value by one.

The material in `\everypar` effectively "cuts in" ahead of the character (or whatever) that made TeX start the paragraph, but it is read already in horizontal mode. This means that if anything inside \everypar makes TₑX go into vertical mode, you get into an endless loop. Just for fun, try saying

$$\texttt{\everypar\{\message\{Boo!\}\par\} x}$$

If you're wondering how you can get a \smallskip before every paragraph, check up \parskip.

evolution in TₑX interfaces, page 8; in printing technology, pages 1–2, 5.

ex Pages 40, 51. A keyword for the ex, one of TₑX's units that depends on the current font.

exceptional entries

with \halign page 107–108, 112–115; with tabbing, pages 123, 128.

executing a command

Pages 165, 175.

\exists Page 132. Math mode only. Produces the symbol ∃.

\exp Page 140. Math mode only. Produces the abbreviation exp, which functions as a large operator with no limits.

expanding a macro

Page 165.

\expandafter

This subtle primitive is not for beginners! It causes the next token to be expanded only after the following one. It turns out to be very useful in certain types of programming; for a simple application, see \uppercase.

exploded view Pages 102, 114.

exponent See superscript.

extended fonts

Pages 29.

extensibility of TₑX

Page 5.

extensible symbols

Pages 33, 35, 144–148.

extra-deep, extra-high

characters, pages 91–92; lines, pages 112, 139.

eye Page 3.

\fam Pages 32–35, 139–140, 143. One of TₑX's integer variables; it refers to the current math font family, which ranges from 0 to 15. When it encounters a letter or digit in math mode (or any variable-family character: see page 131 and \mathchar), TₑX takes note of the family number current at the time, and of the current style. As explained on page 139, this character will be typeset in the font associated with this family and style: if the current family is \fam1 and the the style is display or text, the font selected will be \textfont1; if the style is script, the font will be \scriptfont1, and if the style is scriptscript, the font will be \scriptscriptfont1. But the font selection is made *at the end of the formula*: for example,

$$\texttt{\$\fam1\textfont1=\tenrm a \textfont1=\tenbf a\$}$$

will set both 'a's in font \tenbf.

The value of `\fam` is automatically reset to -1 when TEX enters math mode: this value means that even variable-family characters should be typeset from their intrinsic family, which is 1 for letters and 0 for digits. Therefore letters come out in math italics by default and digits in roman. Only when you explicitly reset the value of `\fam` inside a formula are these conventions changed.

family of math fonts
:   See the previous entry.

`fancy` **format**
:   Page 66. The page layout format defined in the `fancy.tex` file described in chapter 7.

fancy paragraph shapes
:   See `\parshape`.

ff, ffi, ffl, fi
:   Page 19. Ligatures obtained by typing `ff`, `ffi`, `ffl`, `fi`, except with typewriter fonts.

`\fi`
:   Pages 183–185. Used in conditionals: `\if ... \else ... \fi` or `\if ... \fi`.

`fil`
:   Pages 48, 56. The unit of weak springiness, it can be part of the `plus` or `minus` component of glue: for example, `\hfil` is equivalent to `\hskip 0pt plus 1fil`. If you add glue containing `fil` to finite glue, the finite component disappears:

    ```
    \newskip\scratch \scratch=3pt plus 10in
    \advance\scratch by 0pt plus 1fil \showthe\scratch
    ```

    will print `3.0pt plus 1.0fil` on your screen. This is why if a box contains both a spring and finite glue, only the spring will stretch to fill up any available space.

    If an `l` comes after `fil`, even after spaces, it gets incorporated into the keyword, and the result is `fill`. See fill and `\relax`.

`\filbreak`
:   Page 77. Stands for `\par\vfil\penalty-200\vfilneg`. Makes the current spot a good potential page break, even if the current page is not complete; the page will be padded with white space if the break occurs. For concreteness, say the page height is 8 inches, and you have chunks of text (say short sections, program listings, bibliographical entries) that are 2, 7, 3, 3 and 7 inches tall. You want chunks not to be broken across pages if at all possible, but they don't necessarily have to start a fresh page. If you insert `\filbreak` between the chunks, the layout will be as follows: page 1 will get the first chunk only, because the second chunk doens't fit whole in the remaining six inches. Page 2 gets the second chunk; page 3, the third and fourth; and page 4 gets the last chunk.

    As explained under `\everypar`, you can't use that command to insert `\filbreak` between paragraphs automatically. But you can say

    ```
    \let\oldpar\par \def\par{\oldpar\vfil\penalty -200\vfilneg}
    ```

    to make `\par` imitate `\filbreak`. Still, you need to be careful: playing with the definition of `\par` can lead to surprises.

file
:   See source file, `\input`, dvi file, `log` file, fmt file, tfm file.

`fill`
:   Page 48. `fill` is the unit of strong springiness; it is to `fil` as `fil` is to the finite units `cm`, `in`, and so on. This means that if you add glue of type `fill` to glue of type `fil` (or finite glue), the weaker glue is thrown out: see under `fil`. The spring `\hfill` is equivalent to `\hskip 0pt plus 1fill`.

    If an `l` comes after `fill`, even after spaces, it gets incorporated into the keyword, and the result is `filll`. See the next entry and `\relax`.

filll In a pinch you can make super-strong springs using `filll`, which cancels even `fill` (see the previous entry). The use of `filll` is truly exceptional, so TₑX doesn't provide corresponding abbreviations \hfilll and \vfilll.

  If an `l` comes after `filll`, even after spaces, TₑX complaines you're pushing it too far. See \relax.

financial reports

  Page 173.

fine-tuning Pages 87, 92, 131, 148, 153; see also \looseness.

\firstmark The first mark encountered on a page that TₑX has just completed. See \mark.

fixing dimensions

  of a box, page 88; of a table, page 117.

\fivebf, \fiverm

  Page 28. Plain TₑX defines five-point fonts primarily for use in math second-order subscripts (page 35), but these two, boldface and roman, can also be used in text. The other plain TₑX five-point fonts are \fivei (math italics) and \fivesy (math symbols).

fl Page 19. Ligature obtained by typing `fl`, except with typewriter fonts.

\flat Page 132. Math mode only. Produces the symbol ♭.

floats Pages 70, 76; see also insertions.

fmt file File containing a precompiled TₑX format, and read in at high speed: see \dump.

\folio Pages 64–65, 71. Prints the current page number, stored in \pageno. By convention, a negative \pageno is printed as a roman numeral; therefore plain TₑX's definition for \folio is

      \def\folio{\ifnum\pageno<0
        \romannumeral-\pageno \else\number\pageno \fi}

  For uppercase roman numerals, replace \romannumeral by \Romannumeral, whose definition is given in the appropriate entry (it is not part of plain TₑX).

  Normally, \folio is called automatically, by some command that is activated when a page is output. Plain TₑX sets \footline={\hfil\tenrm\folio\hfil}. The `fancy` format of chapter 7 uses the same footline by default, except on the first page, or on a page where you're said \titlepagetrue. You can also say \titlepagefalse to get a folio number on the first page.

  To get no page numbers at all, type \nopagenumbers.

font change, see page 8, \font, \fam; Computer Modern, pages 27–30, 36; current, pages 39–40, 47, 66, 101, 140; Cyrillic, page 30; dimensions, see pages 6, 9, 101, \fontdimen; files, page 9; in footline and headline, page 66; Fraktur ("gothic"), pages 30, 34; metric information, see pages 6, 9, \fontdimen; names, pages 29, 165; in math mode, pages 32, 135; outline, page 31; preloaded, page 27; in plain TₑX, page 28; Postscript, pages 6, 9, 27, 29, 31, 34, 36; quality page 2; registration, page 28.

\font Page 29. Registers a font name: after \font\toto=cmss10, no font change occurs, but the control sequence \toto acquires the meaning "make the font found on file `cmss10` the current text font." Saying \toto will still not change fonts inside a math formula; to do that you must say (for example) \textfont\myfam=\toto \myfam, as explained under \fam. See pages 23, 33–34, 37, 49, 92 for examples, and the entries `at` and `scaled`.

Another use of `\font`, not discussed in the text, is to refer to the current font, in contexts where TEX is already expecting to read a font. For example, `\fontdimen5\font` gives the ex-height for the current font, and `\the\font` generates a control sequence associated with the current font.

Warning: the binding of control sequences to font files is irreversible. After `\font\toto=` cmr10 the construction `\the\font` yields `\toto`, assuming the current font is cmr10. But if you subsequently say `\font\toto=cmbx10`, changing the meaning of `\toto` altogether, TEX will still respond with `\toto` if you say `\the\font` while the current font is cmr10. This is not very sensible, and wouldn't matter much except that TEX uses the same wrong names when describing the contents of a box. So try not to use the same control sequence for different fonts.

`\fontdimen`   Page 101. Among the metric information that TEX needs about its fonts and that it reads from the font's `tfm` file are certain dimension parameters, like the design size, the ex-height, and so on. These parameters can be accessed (and changed) with the construction

$$\text{\texttt{\textbackslash fontdimen}}\ n\ \textit{font}$$

where n is the parameter number and *font* is something like `\tenrm` or `\textfont\bffam`. You can use this construction anywhere like a dimension register, except with `\advance`, `\multiply` and `\divide`.

Natually, direct use of such low-level stuff is not common, but one parameter, the height of the math axis, is sometimes useful to know. It is stored in `\fontdimen22\textfont2`.

`\footline`   Pages 64, 67, 185. A token list from which a page's footline is made when the page is ready to be shipped out: the plain TEX equivalent of `\evenpagefoot`, but used for all pages. Often set to display the page number.

`\footnote`   Pages 61–62. Macro used for footnote insertions. Takes two arguments, the mark and the text of the note; see examples of use on pages 4, 11. Don't leave spaces or a CR before the call to `\footnote`, as it will appear in the output just before the mark; use a `%` if you start want to type `\footnote` on a new line:

```
...word to which the note applies.%
\footnote{(*)}{Text of the note.}
```

Notice also that the note is set in whatever happens to be the current font. For a more sophisticated macro that takes care of these details and also numbers notes automatically, see `\myfootnote`.

`\footnote` won't work inside a vertical box, or inside a horizontal box that is part of a paragraph or box of any kind; the footnote will disappear. See insertions for details.

Other references to footnotes can be found on pages 7, 34, 61, 76, 91.

`\footnoterule`

The rule (and glue) that separates a footnote from the rest of the page. Set as follows:

```
\def\footnoterule{\kern -3pt \hrule width 2truein \kern 2.4pt}
```

In addition, plain TEX leaves the equivalent of a `\bigskip` just before `\footnoterule` is called. You can redefine `\footnoterule` at will; in particular, removing the `width 2truein` will make the rule extend to the width of the page.

foreign languages

Pages 4, 7, 19–20, 27, 30, 174; see also `\discretionary`, `\frenchspacing`, `\hyphen-ation` and `\language`.

`\forall`   Page 132. Math mode only. Produces the symbol ∀.

format file   Pages 5, 64–65; see also `\dump`.

formulas   see math mode (chapter 11); aligning, pages 154, 187; breaking, page 153; numbering, pages 155–156, 189.

Fortran   Page 8.

Fouquet, Nicolas
                     Page 60.

fractions   See `\over`, `/`.

`\frak`   Pages 34–35. Not part of plain T_EX. Switches to a Fraktur ("gothic") font, in text or in math mode. See the next entry.

Fraktur fonts   Pages 30, 34. The American Mathematical Society distributes a set of Fraktur fonts.

frame around a box
                     Page 97.

French typography
                     Pages 20, 174; see also `\oe` and the next entry.

`\frenchspacing`
                     Tells T_EX *not* to follow the conventions of Anglo-Saxon typesetting regarding spacing after punctuation. Ordinarily, T_EX will leave more space after a comma than the ordinary interword spacing, and even more between sentences, that is, after a period, colon, etc.

                     The dot after an abbreviation should not be followed by extra spacing, which is one reason to use a tie instead of a SP in such positions: see `~`. But in bibliographies and other contexts where there are lots of abbreviations it may be easier to say `\frenchspacing` once and for all, so no ties are necessary.

                     For the definition of `\frenchspacing`, see `\spacefactor`. For the opposite behavior, see `\nonfrenchspacing`.

frieze   Page 107.

front ends   Pages 3, 7.

`\frown`   Page 134. Math mode only. Produces the relation ⌢.

function names
                     Pages 140, 143.

`\gamma`, `\Gamma`
                     Page 132, 151. Math mode only. Produce the Greek letter γ, Γ.

Gandalf the Grey
                     Page 11.

`\gcd`   Page 140. Math mode only. Produces the abbreviation gcd, which functions as a large operator, with limits in display style.

`\gdef`   Pages 165. An abbreviation for `\global\def`: used when a definition made inside a group must survive after the group ends. For an example, see page 175.

generalized letter
                     Page 177. Any character of category 11 is considered a "letter" in the scanning of a control sequence name: see escape character for more details. Making a non-letter like @ have category 11 for a while allows one to define control sequences that cannot be accessed or overwritten after the category is changed back.

\ge, \geq   Page 134. Math mode only. Produce the relation $\geq$, for "greater than or equal to." The two control sequences are synonymous: we suppose \geq is for masochists.

German   Page 174; see also \discretionary, \ss.

\gets   Page 134. Math mode only. Produce the relation $\leftarrow$, and is synonymous with \leftarrow.

\gg   Page 134. Math mode only. Produce the relation $\gg$, for "much greater than."

Gibbon, Edward
    Page 108.

\global   Pages 24, 62, 65, 71, 84, 125, 165, 179, 185. Makes an immediately following assignment or definition global, in the sense that it will remain in effect even after the current group ends; ordinarily such changes die with the group. See also \gdef.

global   assignment or definition, see previous entry; magnification, page 31.

glue   Pages 40, 45. Spacing that can stretch or shrink to meet a target. When TEX is expecting a blob of glue (page 181), it will accept either an "internal glue," optionally preceded by a sign, or a specification of the form

$$d_0 \text{ plus } d_+ \text{ minus } d_-$$

An internal glue is a construction of the form \skip n, or any control sequence defined with \newskip, or a glue variable like \parskip; while d_0, d_+ and d_- are arbitrary dimensions (see dimensions), representing the natural, or ideal, dimension for the glue, its stretchability, and its shrinkability. TEX will never shrink glue by more than d_-, but the maximum stretchability can be violated, at a cost: see \hbadness. The plus and/or minus components may be absent, in which case the corresponding dimensions are assumed to be zero; but if present they must come in that order. They may also be negative, and plus -5pt is not the same as minus 5pt: see \section. Finally, they may be infinite (positive or negative), if they involve the units fil, fill or filll rather than "normal" units like in.

Glue is created in many ways: by a space in horizontal mode (page 40), by the \hskip and \vskip commands (pages 39-40) and macros that call these commands, automatically between boxes and lines that are being stacked up (pages 93, 112), between columns of an alignment (pages 117–119), and so on.

Horizontal glue can only be used in horizontal mode and math mode, and it contributes to the width of the enclosing horizontal box or line (page 81) or alignment entry (page 103). Vertical glue can only be used in vertical mode, and it contributes to the vertical dimensions of the enclosing vertical box or page, and it also influences the placement of the baseline (page 86).

If several blobs of glue are present in the same box, they share the available space (or, which is the same, the deficit toward the target box dimension) in proportion to their stretchability or shrinkability: see page 42 and also fil, fill.

TEX will gladly break a line or page at a blob of glue, in which case the glue disappears: pages 44, 68. For unbreakable spaces, see \kern and \penalty; for spacing that doesn't disappear at a break, see \hglue and \vglue.

goals of TEX   Page 2.

God   Page 61.

\goodbreak   Page 77. An abbreviation for \penalty -500. In horizontal mode, it marks the place as a good one for a line break. It doesn't start a new paragraph, however, even if the break occurs. In vertical mode, it marks the place as an excellent one for a page break. Compare \bigbreak, which first puts T_EX in vertical mode, then marks the place as a good page break.

the subsequent ones won't. If `\hangindent` is positive, the indentation applies to the left margin; if it is negative, it applies to the right margin.

The values of `\hangindent` and `\hangafter` that apply to a paragraph are those in effect at the end of the paragraph. If you set `\hangindent` or `\hangafter` inside a group—say, because you're also changing `\parindent` temporarily—you must close the paragraph before the group ends and variables revert to their previous value: see page 54.

The first line of a paragraph with hanging indentation is further indented by `\parindent`, unless you start it with `\noindent`. See also `\displaywidth`.

hanging indentation
:   See the previous entry, and page 55.

`\harr`
:   Pages 161–162. Math mode only. Not part of plain TEX. Draws a horizontal arrow with labels, to be used in diagrams.

hash mark
:   See `#`.

`\hat`
:   Page 137. Math mode only. Places a circumflex accent, or hat, over the following character: $\hat{a}$. Its text counterpart is `\^`. See also `\widehat`.

`\hbadness`
:   One of TEX's integer variables. As explained on page 41, TEX assigns a grade to each line, page or box whose glue had to stretch or shrink to meet a goal. An overfull or underfull box is reported if the badness grade exceeds `\hbadness`, for lines and horizontal boxes, or `\vbadness`, for pages and vertical boxes.

    A badness of 100 means the glue has stretched or shrunk as much as it was designed to: for instance, 12 pt plus 8 pt minus 4 pt has stretched to 20 pt, or shrunk to 8 pt. If the glue has stretched twice as much as it was designed to, in this case to 28 pt, the badness is $100 \times 2^3 = 800$, so the badness grows fast with increasing stretching. In no event will glue shrink by more than its design shrinkability; in this example it would not shrink to less than 8 pt. Instead TEX would report an overfull box, assign it the maximum badness, 10000, and (if a horizontal box) highlight it on the page with a black stroke. But see also `\hfuzz`.

    Plain TEX sets `\hbadness=1000`, which is pretty lax: glue can stretch by more than twice its design stretchability. To set a more stringent standard, lower `\hbadness`; to set a laxer one, increase it.

`\hbar`
:   Page 132. Math mode only. Produces the symbol $\hbar$.

`\hbox`
:   Page 80. Introduces a horizontal box, made by placing side by side each element in its interior. Often used to create a "line" of text that's only as long as the text itself (unlike lines in paragraphs which extend across the page), and to temporarily go out of math mode inside a formula. Examples of use appear on pages 33, 41–42, 47, 51, 59, 80, 84, 88, 96–97, 101, 109, 119, 135, 137–138, 143, 147, 162, 181; see boxes for a breakdown.

`\headline`
:   Pages 64–65, 183. A token list from which a page's headline is made when the page is ready to be shipped out: the plain TEX equivalent of `\evenpagehead`, but used for all pages. Can display the work's title and author, the page number, and so on.

`\heartsuit`
:   Pages 132. Math mode only. Produces the symbol $\heartsuit$.

Hebrew
:   See pages 7, 30, `\aleph`.

height
:   of boxes, see pages 78–80, 89, `\ht`; of characters, page 9; of pages, see `\vsize`; of rules, see next entry. See also dimensions.

`height`
:   Page 94. An optional specification after `\hrule` or `\vrule`; must be followed by a dimension. A horizontal rule without this specification will have height .4 pt; a vertical rule

will be as tall as the immediately enclosing box or line. See examples on pages 115 and following.

help messages
> Page 15; see also error checking and recovery.

Helvetica fonts
> Pages 27, 31.

Henriques, E. Frank
> Page 58.

hexadecimal numbers
> See page 36, integers, `\mathchar` .

`\hfil`
> Pages 42–43. A weak horizontal spring: it stretches to fill the available space in a box, alignment entry, etc., unless there is a stronger spring (see the next entry) in the same box. Several `\hfil`s in the same box share the available space equitably. Although a primitive, `\hfil` is essentially the same as `\hskip 0pt plus 1fil`. See examples of use on pages 45, 48, 56, 105, 141, 150.

`\hfill`
> Pages 42–43. A strong horizontal spring: it stretches to fill the available space in a box, alignment entry, etc., preventing weak springs from stretching altogether. Several `\hfill`s in the same box share the available space equitably. Although a primitive, `\hfill` is essentially the same as `\hskip 0pt plus 1fill`. See examples of use on pages 45, 49, 50, 59, 127, 141, 153, 157, 189.
>
> One common use of `\hfill` is with macros that center their arguments, such as `\matrix`, etc. Such macros use `\hfil` for the centering, so you can counteract that with `\hfill`. For the same reason, use `\hfil` if possible, rather than `\hfill`, when writing your own macros.

`\hfilneg`
> A weak horizontal spring whose stretchability is infinitely negative: although a primitive, `\hfilneg` is equivalent to `\hskip 0pt plus -1fil`. Unlike `\hss`, this spring doesn't shrink to a negative length: its only use is to cancel another, positive, spring placed somewhere else in the same box. An example of the related `\vfilneg` command is given under `\filbreak` .

`\hfq`
> Pages 105, 107. Not part of plain TEX. Abbreviation of `\hfil\quad` ; very useful in alignment entries.

`\hfuzz`
> Page 16. One of TEX's dimension variables, set by plain TEX to .1 pt. An overfull line or horizontal box is not reported if the excess material is less than `\hfuzz`. If you think it's OK for a line to stick, say, 2 pt into the margin, set `\hfuzz=2pt` ; then a line will have to be worse than that before TEX bothers you with it.

`\hglue`
> Pages 44–45. Horizontal "glue" that will not disappear at a line break. It's really an invisible rule followed by normal glue.

`\hidewidth`
> Pages 49, 108, 128. A spring that starts with a negative width and can stretch arbitrarily. Used to hide the width of an alignment entry, so that entry doesn't influence the width of its column. Its effect is to allow the entry to spill into an adjacent column, to either side, depending on which end of the entry `\hidewidth` is at. You can even use `\hidewidth` on both sides, to make the entry will spill over symmetrically: see `\oalign` for an example.

history
> of England, page 108; of France, page 60; of India, page 63; of Rome, page 108; of TEX, page 1; world, page 108.

The Hobbit
> Page 11.

\hom   Page 140. Math mode only. Produces the abbreviation hom, which functions as a large operator with no limits.

\hoffset   One of TEX's dimension variables; it controls the horizontal offset of the text with respect to your sheet of paper. When \hoffset is zero, the default, your printer is supposed to place the left margin one inch from the edge. A positive offset moves the margin to the right, and a negative one to the left. Useful in centering your text when \hsize is different from its default value of 6.5 in, or your paper has width different from 8.5 in.

\hookleftarrow, \hookrightarrow
                Page 134. Math mode only. Produce the relations ↩, ↪.

horizontal   boxes, see boxes, \hbox ; mode, see pages 24, 38, 52, 83 and mode change; rules, see \hrule , \vrule ; spacing, pages 39, 45.

hot lead   Page 1.

house plants   Page 47–48.

\hphantom   Pages 111, 136, 151. Puts its argument inside a horizontal box, measures its width, then typesets an empty box of the same width and zero height and depth. The material in the argument does not appear on the page. See also \phantom , \vphantom , \smash .

\hrule   Pages 94–96. Creates a rule (straight line, or rectangle) of specified dimensions. The full construction is

$$\text{\hrule height } h \text{ depth } d \text{ width } w$$

with the attributes always in that order. If height is missing, it is set to .4 pt; if depth is missing, it is set to zero; if width is missing, the rule is as wide as the immediately enclosing vertical box, or the page.

\hrule must appear in vertical mode, or TEX will switch to it: page 52. But see the next entry for an exception. To draw a horizontal rule in a paragraph, use \vrule : page 97. No glue is placed above or below an \hrule as it gets stacked with other elements in a vertical box or on the page: pages 81–82.

For use in alignments, see pages 109, 121, 125. For other examples of use, see pages 67, 97–98.

\hrulefill   Pages 46, 67, 114, 126. A rule that behaves like a spring, or is it the other way around? Used in horizontal mode, in an \hbox or in a paragraph, it creates a horizontal rule of height .4 pt and depth zero, stretching as far as \hfill would in the same circumstances:

\hbox to 1in{a \hrulefill\ rule} . . . . a ————— rule

The definition of \hrulefill is \def\hrulefill{\leaders\hrule\hfill} . You can specify a height and a depth after \hrule for special effects. (You can also set width , but that has no effect.)

\hsize   Page 13. One of TEX's dimension variables: it controls the width of the page. Plain TEX sets \hsize=6.5in . When TEX encounters any horizontal mode material while in vertical mode (see mode changes), it switches to ordinary horizontal mode and composes a paragraph, which is then chopped up into lines of width \hsize : pages 41, 80, 109, etc. The moral of the story is that if you want short lines of text stacked up in a vertical box, you must fashion each line individually with \hbox , as on page 124, or make \hsize small, as on pages 83, 99.

\hskip   Page 39. Creates horizontal glue of a specified size, with specified stretchability and shrinkability. For more details, see glue.

\hss   Page 49. Horizontal spring that can stretch or shrink indefinitely, taking on a negative width. Useful for things that should be allowed to spill over: see \centerline, \llap, \rlap. For another example, see page 96.

\ht   Pages 88, 181, 183. The construction \ht n gives the height of box n. For details, see \dp.

humanities Page 7.

Hume, David Page 108.

Humpty Dumpty
            Page 171.

Hungarian umlaut
            Page 20.

hyphen   See -, \-.

hyphenation Pages 4, 7, 19, 41, 75, 166. See also \-, \language, \showhyphens and the next entry.

\hyphenation

Adds one or more entries to a dictionary of exceptional words that are not correctly handled by TEX's normal hyphenation rules. Plain TEX says

```
\hyphenation{as-so-ciate as-so-ciates dec-li-na-tion
    oblig-a-tory phil-an-thropic present presents project
    projects reci-procity re-cog-ni-zance ref-or-ma-tion
    ret-ri-bu-tion ta-ble}
```

Each allowed break is indicated by - , not by a discretionary hyphen \- . Words are separated by spaces, and they must consist entirely of letters: no accents are allowed. To add your own entries, just follow the same pattern. You can do it as many times as you want. Once learned, a word is never forgotten, even at the end of a group. See also \showhyphens.

TEX can handle different sets of hyphenation rules for different languages; \hyphenation only affects the hyphenation table for the current value of \language. (This doesn't affect versions of TEX prior to 3.0.)

i, I   Page 15, 26. When TEX has stopped because of an error, typing i or I lets you insert text to help TEX recover. For more details and for other options, see error checking and recovery.

\i   Pages 20–21. Produces a dotless ı, to be used with accents: \^\i, \"\i. Works in text only; in math mode it gives the character ⊂! But there is a math mode equivalent, \imath.

\ialign   Page 186. The many plain TEX macros based on \halign must rely on certain variables having known values: for example, \tabskip should be zero, and \everycr should be empty. The \ialign macro makes these initializations and ends by calling \halign. The moral of the story: to be on the safe side, use \ialign in your macros, instead of \halign, if you think those variables might have been played with.

IBM   Page 5.

\if   One of TEX's more esoteric conditional tests—make sure you understand about tokens before you try to use it! It tests true if the next two tokens found *after macro expansion* are the same character (possibly with different category codes) or if they are both control sequences (possibly different). For example, after \def\beast{aardvark}, saying \if\beast TRUE\else FALSE\fi will print rdvarkTRUE. More usefully, you can make your own boolean variables: \if\toto T will test true if you've previously said \def\toto{T}, but false if you've said \def\toto{F}. (Another way to create boolean variables is with \newif, but this is more economical.)

\ifcase   Page 185. One of TeX's conditional tests. It tests whether the following integer is 0, 1, ..., and branches accordingly. See \today and \magstep for examples of use.

\ifcat   One of TeX's more esoteric conditional tests. It tests true if the next two tokens found *after macro expansion* are (possibly different) characters with the same category code, or if they are both (possibly different) control sequences.

\ifdim   Pages 182, 185. One of TeX's conditional tests. Tests whether the two dimensions that follow, separated by a relation = , < or > , satisfy that relation.

\iff   Page 134. Math mode only. Produces the symbol $\Longleftrightarrow$, with a thick space on each side, unlike \Longleftrightarrow , which is treated as a relation.

\iffalse   One of TeX's conditional tests: always tests false. This may seem silly, but it's very useful in macros. Plain TeX's \newif scheme is based on it: when you say \newif\iftoto , this causes two definitions to take place:

             \def\tototrue{\let\iftoto\iftrue}
             \def\totofalse{\let\iftoto\iffalse}

             The next time you say either \tototrue or \totofalse , the "conditional" \iftoto acquires the desired meaning.

\ifhbox   One of TeX's conditional tests. Followed by a number n between 0 and 255, tests whether box n is a horizontal box.

\ifhmode   One of TeX's conditional tests. Tests true if TeX is in horizontal mode, whether ordinary or strict.

\ifinner   One of TeX's conditional tests. Tests true if TeX is in internal vertical mode, in strict horizontal mode, or in text math mode (between single dollar signs).

\ifmmode   Pages 185. One of TeX's conditional tests. Tests true if TeX is in math mode. Here's a simple but useful application: Suppose you want to abbreviate α to \a. The abbreviation will work fine as long as you want an α by itself, but inside a formula like $\alpha + \beta$, you can't say $\a+\beta$, since this will expand to $$\alpha$+\beta$! No problem: with \ifmmode , you can check whether \a should expand to \alpha or to α :

             \def\a{\ifmmode\alpha\else$\alpha$\fi}

\ifnextchar Not part of plain TeX. A handy control sequence that no experienced user will want to be without. It lets you look ahead and decide what to do based on what the next token is— in particular, it lets you create macros with optional arguments. Consider the following definitions:

             \def\mybox{\ifnextchar[{\myBox}{\hbox to 2in}}
             \def\myBox[#1]{\hbox to #1}

             When TeX sees \mybox , it checks whether or not the next token is a [. If not, the result is \hbox to 2in ; but if it is, the result is \hbox to whatever follows in brackets. (The brackets delimit the argument of \myBox : see page 169.)

             Here's the definition of \ifnextchar . For it to work, the control sequences \tempa to \tempe should not be used otherwise.

             \def\ifnextchar#1#2#3{\let\tempe=#1%
               \def\tempa{#2}\def\tempb{#3}\futurelet\tempc\ifnch}
             \def\ifnch{\ifx\tempc\tempe\let\tempd=\tempa
               \else\let\tempd=\tempb\fi\tempd}

\ifnum   Pages 181, 184. One of TEX's conditional tests. Tests whether the two integers that follow, separated by a relation = , < or > , satisfy that relation. For examples, see page 184 and \folio .

\ifodd   Pages 181, 183, 184. One of TEX's conditional tests. Tests if the following integer is odd. (There is no \ifeven .)

\iftitlepage

     Pages 65, 185. Not part of plain TEX. In the fancy format of chapter 7, decides if the current page should get a headline and footline appropriate to a title page (\titlepagehead , \titlepagefoot) rather than the regular ones. The value of \iftitlepage is controlled with \titlepagetrue and \titlepagefalse .

\iftrue   One of TEX's conditional tests: always tests true. See \iffalse .

\ifvbox   One of TEX's conditional tests. Followed by a number n between 0 and 255, tests whether box n is a vertical box.

\ifvmode   One of TEX's conditional tests. Tests true if TEX is in vertical mode, whether ordinary or strict.

\ifvoid   One of TEX's conditional tests. Followed by a number n between 0 and 255, tests whether box n is void (that is, has never been set or was erased when it was used).

\ifx   One of TEX's more esoteric conditional tests. After \ifx , the next two tokens are compared *without macro expansion*. The test is true if the tokens are "the same" in the following sense: If they're both characters, they must be the same character and have same category code; if they're both macros, they must have the same replacement text; if they're both primitives, they must be the same; and similarly for control sequences defined with \font , \countdef , and so on. Synonyms (control sequences defined with \let) count as being the same as whatever they've been \let equal to. For an example of use, see \ifnextchar .

ignored characters

     Page 172. Any character of category 9 is simply ignored when TEX reads your input file. The only such character in plain TEX is NULL, which has ASCII code 0: there's no reason it should be in your input file at all.

\ignorespaces

     Pages 63, 92, 174. Causes TEX to ignore all space tokens encountered from then on, until the occurrence of a non-space token returns things to their normal state. During this process TEX is expanding macros: it's the tokens resulting from the expansion that count.

     Very useful in writing macros that "clean up" after the user, avoiding spurious spaces. The page references above provide examples.

Ikabruob, Nicolas

     Page 71.

\Im   Page 132. Math mode only. Produces the symbol $\Im$, for the imaginary part of a complex number. But this symbol seems confusing: many mathematicians now prefer Im, which you can define with \def\Im{\mathop{\rm Im}} (cf. page 143).

\imath   Page 132. Math mode only. Produces the symbol $\imath$, a dotless 'i', to be used with math accents: $\vec\imath$ gives $\vec\imath$. Its text counterpart is \i .

implementation dependencies

     Pages 6–8, 10, 15–16, 29, 188; see also CR, \dump , \input , \special .

`in`   Pages 12–13, 32, 39. A keyword for inch, one of TEX's units.

`\in`   Page 134. Math mode only. Produces the relation ∈.

`\indent`   Pages 4, 46, 52, 62. In vertical mode, starts a new paragraph, indenting the first line by an amount given by the `\parindent` variable. Inside a paragraph it also creates the same amount of spacing; thus `\indent\indent` at the beginning of a paragraph causes double indentation. See also `\noindent`.

indentation   negative, pages 53, 55; regular, page 53; hanging, page 55 and `\hangafter`.

indexing   Page 7.

`\inf`   Page 140. Math mode only. Produces the abbreviation inf, which functions as a large operator, with limits in display style.

infinite   recursion, page 174, `\everypar`; stretchability, see springs.

`\infty`   Page 132. Math mode only. Produces the symbol ∞.

inner subformulas

For the purposes of spacing, subformulas of a mathematical formula can be assigned any of the classes 0–6 listed on page 131, as well as an eighth one. The last class consists of inner, or delimited, subformulas, and it includes fractions, `\left...\right` constructions, and anything made with `\mathinner`. An inner subformula is separated from most other things by a thin space (given by `\thinmuskip`), but only in text and display styles. As with any symbol, you can avoid the introduction of spacing around an inner subformula by putting it in braces.

`\input`   Pages 14, 50, 65. Causes TEX to put on hold the reading of the current file and to read the named file. What exactly constitutes a file name is implementation-dependent, but a sequence of letters and/or digits followed by a space should work on all systems. If the given file name does not contain an extension, like `.tex` or `.mac`, TEX will add the extension `.tex` to it before looking for the file.

If you run TEX interactively, it expects the first thing you type to be a file name, as if there were an implicit `\input` before it. To override this, see under `*`.

input lines   See CR, blank lines.

inserting text Page 15.

insertion   An insertion is something that logically belongs at a certain point in the main text, but physically can, or should, appear somewhere else. Plain TEX provides the two commonest kinds of insertions: floats, obtained with `\midinsert` and `\topinsert`, and footnotes. Like all insertions, these won't work except at the "outer level:" if you use `\footnote`, `\midinsert` or `\topinsert` inside a vertical box, or inside a horizontal box that is part of a paragraph or box of any kind, the insertion will disappear. Insertions of the same class always appear in the order of their references in the text, so even if you have a big figure that doesn't fit on the current page followed by a small one that does, TEX will defer both till the next page.

An experienced user can deal with insertions in great generality, defining new types and treating them in different ways. For instance, in this book a new type of insertion was used to typeset the parallel text of section 7.8. We won't go into details here; if you're interested in such wizardry, it's time to read the *The TEXbook*.

inspecting a register

Page 180.

\int   Page 133, 138–139, 142. Math mode only. Produces the large operator $\int$, $\displaystyle\int$. Unlike most large operators, \int has its limits placed in the subscript/superscript position, rather than above and below it. You can change this behavior with \displaylimits.

integers   Pages 178, 181. When T_EX is expecting to read an integer, it will accept either an "internal integer" or an integer in decimal, octal, hexadecimal, or character notation. Any of these may be preceded by a sign. Octal notation is introduced by ' , hexadecimal by " , and character notation by ` , as explained under the entries for each of these characters.

interaction with T_EX

See page 26, error checking and recovery, * , \message , \show and subsequent entries.

interactivity   lack of, page 7; levels of, see \batchmode , \errorstopmode , \noscrollmode , and \quietmode .

interbox and interline spacing

automatic, pages 82, 93, 105; changing the, see \openup , \normalbaselines ; with different fonts, see page 35, \eightpoint , \tenpoint ; and rules, page 109.

\interior   Math mode only. Not part of plain T_EX. Places the "interior operator" of topology above the next character: $\interior A$ gives $\mathring{A}$. Its definition is

$$\def\interior\{\mathaccent "7017 \}$$

(the "7017 indicates the position of the symbol in the math fonts, as explained under \mathcode).

internal vertical mode

Pages 25, 79.

interrow spacing

See interbox and interline spacing.

interrupt character

Page 15.

interword spacing

Pages 40, 47; see also \frenchspacing , \spacefactor .

invalid characters

Page 172. Any character of category 15 in your input file causes T_EX to issue an error message. The only such character in plain T_EX is DEL, which has ASCII code 255: there's no reason it should be in your input file at all. But this helps detect when you've accidentally given T_EX a file not meant for its eyes.

invisible   box, see empty box, strut, \phantom ; characters, see SP, CR, TAB; delimiters, page 147; digit, page 111; text, page 151. See also \phantom .

\iota   Page 132. Math mode only. Produces the Greek letter ι.

\it   Pages 28, 33–37, 135, 165. Switches to an italic font, in text or in math mode; but in math mode subscripts and superscripts don't work (page 34). This is generally not a problem, since math mode uses a special italic font by default.

Normally \it should be used inside a group, so its effect goes away when the group ends. In plain T_EX \it always switches to the text font \tenit , and to the math family \itfam . To set things up so that \it switches to an italic font in the current size, see \eightpoint and \tenpoint .

italic   correction, page 37, 136, 144–145; fonts, pages 27–29, 135; Greek capitals, page 136.

\item, \itemitem

Pages 58–59, 62–63. Each of these two macros starts a new paragraph and indents the whole paragraph by \parindent or 2\parindent, respectively. The macro's argument is written on the first line's indentation, separated from the text by half a quad. Plain TEX doesn't have an \itemitemitem macro, but one is defined on page 62.

If you're working with a zero \parindent, as in this book, \item won't behave sensibly. To get any mileage out of it, you must increase the \parindent inside a group. In that case you must close the paragraph before the group ends and variables revert to their previous value: see page 54.

As an alternative for \item, see also \meti.

\itfam

Page 34–36. A name for the italic font family to be used in math mode. To select that family, say \fam\itfam. (The \it command does this.) In plain TEX this family only has a \textfont, so subscripts and superscripts don't work.

isotopes

Page 136.

iteration

Page 183.

\j

Pages 20–21. Produces a dotless ȷ, to be used with accents. Works in text only; in math mode it gives the character ⊃! But there is a math mode equivalent, \jmath.

Jekyll and Hyde

Page 173.

\jmath

Page 132. Math mode only. Produces the symbol $\jmath$, a dotless 'j', to be used with math accents: $\vec\jmath$ gives $\vec\jmath$. Its text counterpart is \j.

\joinrel

Pages 146, 150, 162. Math mode only. Introduces a negative kern of 3 math units, or negative thin space. Used to combine various arrows and other mathematical symbols into one. Plain TEX defines \longrightarrow, which gives $\longrightarrow$, as follows:

\def\longrightarrow{\relbar\joinrel\rightarrow}

justification

Pages 4, 10, 40, 55, 96.

\kappa

Page 132. Math mode only. Produces the Greek letter κ.

Kendall, Paul Murray

Page 60.

\ker

Page 140. Math mode only. Produces the abbreviation ker, which functions as a large operator without limits.

kern

Pages 4, 44, 81. Spacing that does not stretch or shrink, and at which no line or page breaks are allowed (contrast with glue). It can be positive, to separate things, or negative, to bring them together. TEX inserts kerns automatically between certain letters:

VA, \hbox{V}\hbox{A} .... VA, V A

For explicit kerns, see the next entry.

\kern

Pages 44–45, 81, 138, 181–182. Introduces a horizontal or vertical kern of a specified dimension, depending on the current mode. In math mode, the kern is horizontal. \kern is followed by a dimension, not by glue: if you say \kern 10pt plus 2pt minus 2pt you'll get a kern of 10 pt and the words plus 2pt minus 2pt in the output (compare page 194). For examples of use, see pages 51, 92, 96–97, 162, 189. See also \mkern.

keywords

Words that have special meaning to TEX but are not preceded by a backslash. Their special meaning is activated only in certain special situations: for example, units such as in or pt are recognized as keywords in situations when T_EX is reading a dimension. The keywords are: `at` and `scaled` (used in registering fonts); `bp`, `cc`, `cm`, `dd`, `em`, `ex`, `in`, `mm`, `mu`, `pc` and `sp` (units); `by` (used in register arithmetic); `depth`, `height` and `width` (rule specifications); `fil`, `fill` and `filll` (units of infinite glue); `minus` and `plus` (used to introduce the elastic components of glue); `spread` and `to` (used in fixing the dimensions of a box); and `true` (preceding a unit).

Each letter in a keyword can be written in upper or lowercase. In addition, `fill` and `filll` are special, in that spaces are allowed between the 'l's.

killing T_EX   Pages 15–16.

Knuth, Donald E.

Pages 1, 4, 5–6, 8, 27, 140, 176.

\l, \L   Page 19. Produce the letter ł, Ł.

labeling formulas

Pages 155–156, 189.

\lambda, \Lambda

Page 132. Math mode only. Produce the Greek letter λ, Λ.

\land   Pages 133. Math mode only. Produces the binary operator $\wedge$, also obtained with `\wedge`.

\langle   Pages 135. Math mode only. Produces left angle brackets $\langle$, not to be confused with the less-than sign $<$. As a math delimiter, it can grow arbitrarily large with `\big`, `\left`, and so on (pages 146–147); in this context you can use $<$ as an abbreviation, since the less-than sign doesn't grow. See also `\lfq`.

\language   One of T_EX's integer variables (not available before version 3.0). It tells T_EX which hyphenation table it should use, assuming more than one such table was preloaded.

Plain T_EX only has a hyphenation table for English, but the typical site in Canada, for example, might support both French and English. Normally such installations will offer mnemonic commands to hide from the user the actual numerical values of `\language`: for example, `\french` might be defined as `\def\french{\language=1\frenchspacing}` and `\english` as `\def\english{\language=0\nofrenchspacing}`.

large operators

Pages 131–132, 139, 142. A math symbol or subformula of class 1. A large operator is separated from a preceding or following symbol of class 0 (ordinary) or 1 (operator) by a thin space (glue given by `\thinmuskip`). Function names are considered "large" operators.

The positioning of a large operator's limits (subscripts and superscripts) depends on style, and this dependence can be modified by the use of `\limits`, `\displaylimits` and `\nolimits`.

If a large operator consists of a single symbol, it is vertically centered with respect to the math axis, and its italic correction is taken into account in the placement of limits.

As with any symbol, you can deprive a large operator of spacing and other special behavior by placing it within braces.

L^AT_EX   Pages 5, 24, 102, 163; see also `\begin`.

layered structure

Page 5.

\lbrace   Pages 135. Math mode only. Produces left braces $\{$, just like `\{`. As a math delimiter, it can grow arbitrarily large with `\bigl`, `\left`, and so on (pages 146–147).

`\lbrack`   Pages 135. Math mode only. Produces left brackets [, just like [. As a math delimiter, it can grow arbitrarily large with `\bigl`, `\left`, and so on (pages 146–147).

`\lceil`   Pages 135. Math mode only. Produces the symbol ⌈, which, besides denoting the ceiling function, is used by plain TEX as the top portion of an opening bracket. As a math delimiter, `\lceil` it can grow arbitrarily large with `\bigl`, `\left`, and so on (pages 146–147).

`\ldisplaylinesno`

Pages 189. Not part of plain TEX. Allows labeling (on the left) of several centered formulas in a display.

`\ldotp`   Math mode only. A period with a bit of spacing before and after. Its direct use is very rare, but `\ldotp\ldotp\ldotp` is how plain TEX defines the next macro, which is used all the time.

`\ldots`   Pages 145, 160, 167 Math mode only. Produces three dots at the right height and with the right spacing to be placed between commas or letters. Gives better results than $\ldots$:

`$x=(x_1,\ldots,x_n)$` .... $x = (x_1, \ldots, x_n)$

`$x=(x_1,...,x_n)$` .... $x = (x_1, ..., x_n)$

`\le`   Page 134. Math mode only. Produces the relation $\leq$, for "less than or equal to." Synonymous with `\leq`.

leaders   Page 46–48. Dots in a row leading the eye across the page, as in an index entry. See also the next entry.

`\leaders`   Pages 47–48, 51, 120. Produces "generalized leaders," a repeated pattern that fills up space as if it were glue or springs. The general syntax is

`\leaders` *prototype glue*

The prototype can be a box or a rule: see `\hrulefill` for an example of the latter. Horizontal glue is expected in horizontal mode, and the prototype is repeated horizontally; vertical glue in expected in vertical mode, and the prototype is repeated vertically. Here's how to set the pattern of page 51 using vertical leaders:

```
\vbox to .5in{
    \leaders\line{\strut\hskip 3cm\multitex\hskip 3cm}\vfill}
```

leading   See pages 40, 52, interbox and interline spacing.

`\leavevmode` Page 52. In vertical mode, starts a paragraph, or, which is the same, switches to horizontal mode. Necessary when you want to start a paragraph with a box or space or something else that can occur in either mode. There are examples on pages 46, 100, 176, but here's another one: if you try to start a paragraph with

`\hbox to 1cm{$\bullet$\hfill} Once upon a time...`

you're in for a surprise:

●

Once upon a time...

One solution is to precede `\hbox` with `\leavevmode`. In this case you could also use `\noindent` or `\indent`, but inside a macro that can occur in the middle of a paragraph as well you wouldn't want to do that.

Le Brun, Charles

Page 60.

`\left`   Pages 147–148. Math mode only. Followed by any delimiter, causes TEX to look for a matching `\right` delimiter, and to make the delimiters as big as necessary to enclose the formula between them. The whole construction is treated as an inner subformula for the purposes of spacing.

The matching delimiter can be a dummy one, represented by `\right.`, but it must be there: see examples on pages 90, 155–156, 158. You can use a closing delimiter with `\left`. For cases where TEX's choice of delimiter size is poor and should be overridden by hand, see page 148.

left delimiters   See opening delimiters.

\leftarrow, \Leftarrow
Page 134. Math mode only. Produce the relations ←, ⇐; the first is also given by `\gets`. See also the next entry.

\leftarrowfill
Pages 46, 117. A "spring" that makes an arrow pointing left:

`\hbox to 1in{\leftarrowfill}` . . . . ←———————————

You can combine it with `\rightarrowfill` to make an arrow that points both ways: `\hbox to 1in{\leftarrowfill\rightarrowfill}`. See also `\overleftarrow` and `\mathord`.

\leftharpoondown, \leftharpoonup
Page 134. Math mode only. Produce the relations ↽, ↼.

\leftitem   Page 59. Not part of plain TEX. Like `\item`, but places the tag flush against the left margin.

\leftline   Page 68. Creates a line containing the material that follows in braces, flush left. It works by creating a box the same width as the page, so it must be used in vertical mode. If you use it inside a paragraph, you'll get an overfull box.

\leftrightarrow, \Leftrightarrow
Page 134. Math mode only. Produce the relations ↔, ⇔. See also `\leftarrowfill`.

\leftskip   Pages 54–55, 57, 167. One of TEX's glue variables. It controls the amount of glue placed at the beginning of each line of a paragraph, and its normal value is zero. Many applications derive from giving it other values: indenting a chunk of text (pages 54, 57), centering each line of a paragraph (page 57), and so on. If you use it inside a group, it's essential to end the paragraph before closing the group, since the value of `\leftskip` used for a whole paragraph is the one in effect at the end of the paragraph (page 54). See also `\displayindent`.

legends   Page 90.

Le Nôtre, André
Page 60.

\leq   Page 134. Math mode only. Produces the relation ≤, for "less than or equal to." Synonymous with `\le`.

\leqalignno   Pages 156, 159, 189. Allows labeling (on the left) of several aligned formulas in a display.

\leqno   Pages 12, 155. Allows labeling (on the left) of a single centered formula in a display.

\let   Pages 166, 173, 176. Saying `\let\foo=\bar` gives `\foo` the current meaning of `\bar` (it "clones" `\bar`). Most often used for abbreviations: after

                    `\let\toto=\longleftrightarrow`

you can use `\toto` wherever you would use `\longleftrightarrow`. The syntax is that of an assignment, and different from that of `\def`; the = is optional. Like all assignments, its effect is restricted to the current group, unless it's preceded by `\global`.

Sometimes \let is the only way to achieve a tricky effect: \bgroup is one such case. Here's another: the definition of \obeylines includes the assignment \let^^M=\par (recall that ^^M is a visible way to represent a CR). If you say \let\par=\cr \obeylines \halign{...}, the effect is that \par first acquires the meaning of \cr, then ^^M acquires the meaning of \par, which is also the meaning of \cr. This means that inside the \halign you can separate rows using a CR, rather than having to type \cr! We used this in the macro that typesets displayed chunks of code in this book (see \everycr).

A clone of an \outer or \long macro is also \outer or \long.

\lfq   Not part of plain TeX. Left "foreign" quotation marks, à la française:

Elle s'est mise à crier: «Au secours! à l'assassin!»

```
Elle s'est mise \`a crier: \lfq Au secours! \`a l'assassin!\rfq
```

Here is its definition, together with that of the matching right quotes:

```
\def\lfq{\leavevmode\raise.3ex\hbox{$\scriptscriptstyle
                                    \langle\!\langle$}}
\def\rfq{\leavevmode\raise.3ex\hbox{$\scriptscriptstyle
                                    \,\rangle\!\rangle$}}
```

\lfloor   Pages 135. Math mode only. Produces the symbol ⌊, which, besides denoting the floor function, is used by plain TeX as the bottom portion of an opening bracket. As a math delimiter, \lfloor it can grow arbitrarily large with \bigl, \left, and so on (pages 146–147).

\lg   Page 140. Math mode only. Produces the abbreviation lg, which functions as a large operator with no limits.

\lhook   Pages 134, 162. Math mode only. Produces the relation ⸜. Mostly used with \joinrel to make longer symbols: \hookleftarrow is composed of (you guessed it!) \lhook and \leftarrow.

ligatures   Pages 4, 9, 19, 81. To prevent a ligature from occurring, it's enough to separate its characters with an empty group: therefore you can write '{}'' for British-style nested quotes, and \tt ?{}' if you program listing contains the string ¿.

\lim   Pages 140, 150. Math mode only. Produces the abbreviation lim, which functions as a large operator, with limits in display style. (Yes, "limit" is an overloaded word...)

\limind   Not part of plain TeX. Math mode only. Two versions of this macro were proposed on page 150: one gives lim ind, the other gives $\underrightarrow{\lim}$. Both function as large operators, with limits in display style.

\liminf   Pages 140. Math mode only. Produces the abbreviation lim inf, which functions as a large operator, with limits in display style. Another version of this macro was proposed on page 150: it gives $\underline{\lim}$.

\limits Pages 142, 162. Math mode only. After a large operator (page 133), declares that limits (or subscripts and superscripts) should be placed above and below the operator in all styles. By default, such operators get limits only in display style.

\limproj Not part of plain TEX. Math mode only. Two versions of this macro were proposed on page 150: one gives lim proj, the other gives $\varprojlim$. Both function as large operators, with limits in display style.

\limsup Pages 140. Math mode only. Produces the abbreviation lim sup, which functions as a large operator, with limits in display style. Another version of this macro was proposed on page 150: it gives $\varlimsup$.

line blank, pages 38, 52, 54, 171; breaks, pages 41, 44, 53, 70; of input, see CR; see also rules, paragraphs, and the next entry.

\line Page 42. An abbreviation for \hbox to hsize. Creates a line, or box the width of the page, containing the material that follows in braces. It doesn't add any springs to the contents, so if they don't have enough elasticity, you'll get an overfull or underfull box. \line should be used in vertical mode. If you use it inside a paragraph, you'll get an overfull box. For examples, see pages 15, 43, 49, 68, 71.

linear algebra Pages 13, 71.

\lineskip, \lineskiplimit
  Page 93. Two of TEX's variables: \lineskip is a glue variable that controls the glue TEX puts between consecutive boxes or lines of text that are being stacked vertically, when the boxes are unusually tall or deep. More precisely, TEX first tries to space them so their baselines are at a distance \baselineskip from one another (see under that entry); but if that would make the glue between the boxes less than \lineskiplimit, which is a dimension variable, TEX uses \lineskip instead. Plain TEX sets \lineskip=1pt and \lineskiplimit=0pt. See \baselineskip for more information.

list formatting, see \item, \meti; of tokens, see token list register.

\ll Page 134. Math mode only. Produce the relation $\ll$, for "much less than."

\llap Pages 48. Backtracks over existing text, then prints its argument. The argument is read in horizontal mode, even if \llap is being used in math mode. See examples on pages 63, 67, 79, 98, 189.

\ln Page 140. Math mode only. Produces the abbreviation ln, which functions as a large operator with no limits.

\lnot Page 132. Math mode only. Produces the symbol $\neg$, also called \neg.

local changes Page 23; see also groups.

log file Pages 14–15, 180. File where TEX places a log of a run, including numbers of completed pages (see \count), error messages, and so on. The name of the file is inherited from the first input file read by TEX: for instance, tex hobbit will read hobbit.tex and create hobbit.log. If the input is interactive, the output goes to texput.log. See also \show and subsequent entries, \tracingcommands and subsequent entries.

\log Page 140. Math mode only. Produces the abbreviation log, which functions as a large operator with no limits.
  entrylogo See \TeX.

\long Page 171. Prefix placed before \def when the arguments of the macro being defined are allowed to contain more than one paragraph. If a macro that is not \long (which is the case with most) is passed an argument containing \par , TEX assumes a mistake was made somewhere and complains of a runaway argument.

long formulas   See pages 142, 153, * , \allowbreak .

\longleftarrow, \Longleftarrow
Page 134, 150. Math mode only. Produce the relations ⟵ , ⟸ .

\longleftrightarrow, \Longleftrightarrow
Page 134, 166. Math mode only. Produce the relations ⟷ , ⟺ . The second symbol is also given by \iff , but then it gets extra spacing around it.

\longmapsto Page 134. Math mode only. Produces the relation ⟼ .

\longrightarrow, \Longrightarrow
Page 134. Math mode only. Produce the relations ⟶ , ⟹ . The former is used in the \showbox macro of pages 98, 101. The latter is used on pages 130, 149.

loop   See page 183, recursion and the next entry.

\loop   The construction \loop *text1* *test* *text2* \repeat provides a basic iteration capability. The *test* is any conditional test (see \if and subsequent entries), without a matching \fi . TEX starts by processing *text1*. It then tests the condition; if it is false, it jumps till after the \repeat , but if it is true, *text2* is processed and the loop starts over. As an illustration, here is plain TEX's definition of \multispan . It makes \multispan3 , for example, expand to \omit\span\omit\span\omit .

\newcount\mscount
\def\multispan#1{\omit\mscount=#1 \loop\ifnum\mscount>1 \sp@n\repeat}
\def\sp@n{\span\omit\advance\mscount\m@ne}

\looseness One of TEX's integer variables. Its normal value is zero, and it is reset to zero after each paragraph. If you set \looseness=1 , TEX will try to loosen the current paragraph so it has one more line than the number it would have otherwise. All resulting lines should still have acceptable badness. Naturally, TEX will only succeed if the paragraph is fairly long or if the last line was almost full. You can also set \looseness to −1 to compress the paragraph, or to 2 to stretch it even more, and so on.

All of that can help at the last stage of page makeup, if, say, a page break would be much better if only you could squeeze another line in. But don't expect miracles.

\lor   Pages 133. Math mode only. Produces the binary operator ∨, also obtained with \vee .

Louis Pasteur University
Pages 68, 71.

Louis XI   Page 60.

Louis XIV   Page 60.

\lower   Pages 87, 181. The construction

$$\text{\lower } dimension \; box \; command$$

has the same effect as *box command*, except that it must occur in horizontal mode, and the box is moved vertically so its baseline is displaced by *dimension* relative to the enclosing box's baseline, instead of coinciding with it. If *dimension* is positive, the box is lowered; if negative, the box is raised. The *box command* can be built on the fly, with \hbox , \vbox , etc., or it can be fetched from memory, with \box or \copy . (But the unboxing commands don't make sense here: the box has to be intact.) For examples, see pages 51, 98, 101.

\lowercase Transforms the characters that follow in braces into lowercase. This is a subtle command: for advice and examples see \uppercase .

\lq A synonym for ' , the left quote. Useful mostly to the poor souls whose keyboards lack the real thing.

Macintosh Pages 5, 7.

macro Pages 6, 51, 164 and following; arguments, pages 58, 167–171; in disguise, page 173; expansion, see pages 15, 165, \expandafter , \if , \ifcat , \ifx , \ignorespaces , \the ; file, pages 14, 16.

macron Page 20.

\magnification
  Pages 31, 39. Sets the global magnification factor for a document; a value of 1000 corresponds to no magnification. This factor affects all dimensions in the document, except those that are given in true units. But it is more accurate to say that the non-true dimensions are unaffected, while the true ones are divided by the magnification factor; the driver program is in charge of actually scaling everything up.

  As a result, a document's line and page breaks don't depend on the magnification unless a true dimension is being used somewhere. Plain TEX does make the \hsize and \vsize true dimensions at the time \magnification is used, in order for the physical size of the page to be the same; so naturally each line fits fewer characters and each page fewer lines.

  It follows also that when a true dimension is specified, TEX is committed to the current magnification; so you can say \magnification only once, before any true dimensions are specified. Furthermore, you must say it before any page is completed.

\magstep, \magstephalf
  Page 31. \magstep1 equals 1200 , so that \magnification=\magstep1 blows things up by a factor of 1.2, or 20%. \magstep2 blows them up again by 1.2, for a total of $1.2 \times 1.2 = 1.44$, or 44%, and so on in multiplicative steps up to \magstep5 . Finally, \magstephalf is the same as 1095 ; the idea is that two half steps multiply up to one step: $1.095 \times 1.095 = 1.2$.

  The definition of \magstep is a textbook example of the use of the conditional \ifcase :

```
\def\magstep#1{\ifcase#1 1000\or
    1200\or 1440\or 1728\or 2074\or 2488\fi\relax}
```

man is a thinking reed
  Page 92.

\mapsto Page 134. Math mode only. Produces the relation ↦.

\mapstochar Math mode only. Produces a little vertical stroke ׀, normally used only in combination with arrows: \mapsto is defined as \mapstochar\rightarrow .

margins Pages 23, 54, 57–58; see also \hoffset , \voffset . To typeset marginal notes, see \marginnote below.

\marginnote Not part of plain TEX. The command defined here can be used to typeset marginal notes; it is not entirely robust, however.

```
\def\marginnote#1{\setbox0=\vtop{\hsize 4.75pc
    \eightpoint\rightskip=.5pc plus 1.5pc #1}\leavevmode
    \vadjust{\dimen0=\dp0
    \kern-\ht0\hbox{\kern-4.75pc\box0}\kern-\dimen0}}
```

We wrote `\marginnote {...}` right after "places," and here's what came out.

This definition places the note 4.75 pc to the left of the main text, since that's the indentation of section headers in this book. You can change that value at will. The `\rightskip` creates a buffer zone between the note and the text; the vertical kerns ensure that the baseline of the note's first line coincides with the baseline of the main line (notice that we must save the value of `\dp1` because it becomes zero after the box is used); and `\vadjust` allows the placement of the note without disturbing the paragraph.

To make the note come out on the right margin, change `\hbox{\kern-4.75pc\box1}` into

$$\hbox{\kern\hsize\kern1pc\box1}$$

Maria Code   Page 59.

`\mark`   Places a mark in the current list of boxes and lines to be printed. The mark itself is not printed, but the first and last marks placed on a page that TEX has just completed are accessible under the names `\firstmark` and `\botmark`. In addition, `\topmark` contains the previous page's `\botmark`.

More exactly, suppose that your file has the commands `\mark{Matthew}`, `\mark{Mark}`, `\mark{Luke}` and `\mark{John}`, and that page breaks are such that the first mark falls on page 2, while the others fall on page 4. By the time page 1 is completed, no marks have been found, so `\topmark`, `\firstmark` and `\botmark` all expand to nothing. At the end of page 2, `\firstmark` and `\botmark` expand to Matthew, but `\topmark` still expands to nothing. At the end of page 3, all three marks give Matthew. At the end of page 4, `\topmark` is still Matthew, `\firstmark` is Mark, and `\botmark` is John. For all subsequent pages, all three marks give John.

The headline of each page of a dictionary often contains the range of entries on the page, and can be built from `\firstmark` and `\botmark`. This Dictionary-Index is no exception: each new entry is introduced by the `\entry` command, whose definition includes an automatic mark: `\def\entry#1{\medbreak...\mark{#1}...}`. The running heads make use of this information in the following way:

```
\evenpagehead={\hbox to 3em{\folio\hfil}\the\runningtitle
    \hfil{\firstmark}\quad--\quad{\botmark}}
```

and similarly for `\oddpagehead`. Note that `\firstmark` and `\botmark` are inside a group, as they may contain font-change commands. Also, in the definition of `\entry`, it is essential that no page break intervene between the mark and the beginning of the entry; if we had written `\def\entry#1{\mark{#1}\medbreak...}` and a break occurred at the `\medbreak`, the last mark on the previous page would be the first entry of the next, and the headers would come out wrong.

In other applications, `\firstmark` may not be adequate to reflect the state of affairs at the top of a page, and one must use instead `\topmark`, or a combination of the two. Such sophisticated mark management won't be discussed here: see pages 258–262 of *The TEXbook*.

A few caveats are in order. First, `\mark` expands its argument, so if the value of a macro is changed between the time it is used in a `\mark` and the time the mark is printed, the change is not reflected in the mark. Next, `\mark` won't work except at the "outer level:" if you use it inside a vertical box, or inside a horizontal box that is part of a paragraph or box of any kind, the mark will never appear as a `\firstmark` or `\botmark` or `\topmark`. Finally, the values of `\firstmark`, `\botmark` and `\topmark` are global, that is, they are not affected by groups.

markup   Page 3.

master file   Page 13.

Masters   Page 102.

mastication   See lexical analysis.

math   article database, page 4; characters, page 131–135; classes of symbols, page 131; font fam-
                 ilies, pages 32–35, 139–140, 143; formulas, see math mode; glue, pages 178, 182; italic
                 fonts, page 135; mode, pages 15–17, 19, 24, 28, 38, 52, 81–82, 84, 91, 100, chapter 11; shift
                 character, see $; symbols, page 17, 33, 131–135; unit, see page 140, mu .

\mathaccent   Math mode only. Primitive from which macros like \hat , \tilde , etc., are built. For an
                 example of use, see \interior .

\mathbin   Math mode only. Makes the following character or group into a binary operator, for the
                 purposes of spacing:

                 $a\tau b$, $a\mathbin{\tau} b$ .... $a\tau b,\, a\,\tau\,b$

\mathchar   Page 36. Math mode only. Followed by an integer $0 \le n < 32768$, prints a character
                 in one of the math font families. The character depends on n in the following way: if
                 $n = 16 \times 256\,x + 256y + z$, with $0 < x < 8, 0 < y < 16$ and $0 < z < 256$ (that is, if x is
                 the highest digit of n in hexadecimal notation, y is the second-highest, and z represents the
                 two lowest), T_EX typesets the character in position z of family y (see \fam), and considers
                 it to be of class x for the purposes of spacing (page 131). The values of x, y and z can be
                 read off easily if n is written in hexadecimal, which is indicated by a " . Here are some
                 examples:

                 • Plain T_EX's definition for \colon is, in essence, \def\colon{\mathchar"603A} ;
                 the 6 indicates a punctuation mark, the 0 says it's taken from font family 0, which is the
                 roman family (page 33), and 3A is the colon's ASCII code in hexadecimal.

                 • Plain T_EX also says \def\alpha{\mathchar"010B} ; here 0 is the class of ordinary
                 characters, 1 is the family of italic fonts, where Greek letters are to be found, and 0B is
                 the position of α in those fonts.

                 • Suppose we use \newfam to define a new family of symbol fonts, as explained on
                 page 34; let its symbolic name be \myfam . To access the symbol that is in position "40 ,
                 say, and treat it as a binary operator, we must type \mathchar"2y40 , where y stands for
                 the hexadecimal digit that T_EX associates with \myfam . We could look in the log file to
                 find out that number, but it might change from run to run; it's much better to deal only with
                 the name, not the number. How can we do that? Writing \myfam in place of y won't
                 work, because this control sequence is an integer register, not a digit. What we need is a
                 way to generate the hex representation of a number, just as \number generates the decimal
                 representation. T_EX provides no such command, but we can easily define one for numbers
                 up to 15:

                     \def\hexnumber@#1{\ifcase#1 0\or1\or2\or3\or4\or
                       5\or6\or7\or8\or9\or A\or B\or C\or D\or E\or F\fi}

                 We can now type \mathchar"\hexnumber\myfam 40 .

                 One last thing: as mentioned on page 131, a character of class 7 is a variable-family charac-
                 ter. This means that the character's family is replaced by the current value of \fam , if \fam
                 was set *within the current formula* to a number between 0 and 15. Otherwise characters of
                 class 7 behave like those of class 0.

                 See also the closely related \mathcode .

\mathchardef
                 Page 36. Makes the following control sequence an abbreviation for a \mathchar con-

struction. It works in the same way as \chardef. The actual definitions of \alpha and \colon (see previous entry) are

$$\text{\texttt{\textbackslash mathchardef\textbackslash alpha="010B \quad \textbackslash mathchardef\textbackslash colon="603A}}$$

\mathchoice   Pages 143, 150. Math mode only. Chooses among the four following groups, depending on the style. Mostly used in macros: after \def\toto{\mathchoice{1}{2}{3}{4}}, you get the following bizarre behavior:

$$\text{\texttt{\$\textbackslash toto+\textbackslash displaystyle\textbackslash toto\textasciicircum\{\textbackslash toto\_\{\textbackslash toto\}\}\$}} \dots\dots\dots\dots\dots\dots 2 + 1^{34}$$

A more serious use is provided by plain TₑX's \root command.

\mathclose   Math mode only. Makes the following character or group into a closing delimiter. The definition of \bigl is \mathclose\big.

\mathcode   Associates with an input character a character in one of the math font families. The input character is referred to by its ASCII code (generally with the '\ mechanism), and the output character by means of the convention explained under \mathchar. For example, plain TₑX says \mathcode'\:="303A; this means that when a : is seen in math mode TₑX prints the character in position "3A of the appropriate font of family 0, and treats it as a relation (class 3). Therefore, \colon and : print the same symbol, but treat it differently for purposes of spacing.

The mathcode of a character can also be "8000, in which case the character is treated as an active character whenever it is encountered in math mode. Plain TₑX sets

$$\text{\texttt{\textbackslash mathcode'\textbackslash'="8000}}$$

and then specifies what ' should expand to:

$$\text{\texttt{\{\textbackslash catcode'\textbackslash'=\textbackslash active \textbackslash gdef'\{\textasciicircum\textbackslash prime\}\}}}$$

(the actual definition is more complicated in order to allow for repeated primes).

mathematical, mathematics
    See math.

\mathinner   Page 160. Math mode only. Makes the following character or group into an inner subformula (see entry). Used very rarely.

\mathop   Pages 143, 149–150, 162. Math mode only. Makes the following character or group into a "large" operator, for the purposes of spacing:

$$\text{\texttt{\$\{\textbackslash rm sin\} x\$, \$\textbackslash mathop\{\textbackslash rm sin\} x\$}} \dots\dots\dots\dots\dots\dots \sin\!x, \sin x$$

\mathopen   Math mode only. Makes the following character or group into an opening delimiter. The definition of \bigl is \mathopen\big.

\mathord   Math mode only. Makes the following character or group into an ordinary subformula, for the purposes of spacing. Plain TₑX uses − and \leftarrow to build up its stretchable arrow \leftarrowfill; it first takes the precaution of making those two characters ordinary, so no spacing is placed around them. (But enclosing a character or subformula in braces is sufficient to make it ordinary.)

\mathrel   Page 162. Makes the following character or group into a relation, for the purposes of spacing. The definition of \bigm is \mathrel\big; other examples were given on page 162.

\mathstrut   Page 151. Produces an invisible rule of width zero and same height and depth as parentheses, used to uniformize the height and depth of different expressions. For examples, see pages 141, 144–145. In spite of its name, \mathstrut can be used outside math mode.

\mathsurround

Page 186. One of TₑX's dimension varibles; it controls the amount of spacing placed before and after a text math expression (one between single dollar signs). Its normal value is zero.

\matrix

Pages 157–158. Typesets a matrix, or array of math formulas. Each entry is centered in its column. Entries are separated by ampersands & , and rows are terminated by \cr ; for examples, see pages 22, 43, 157–158. Page 159 shows how you can increase the spacing between rows of a matrix, and pages 161–162 how \matrix can be adapted to do diagrams. Matrices in parentheses can be obtained with \pmatrix . To surround matrices with other delimiters, use \left...\right (pages 147). Don't use \matrix for systems of equations and such: see pages 186–188 for better solutions. And for a small matrix like $\left(\begin{smallmatrix} a & c \\ b & d \end{smallmatrix}\right)$, see \atop .

\max

Pages 140, 143. Math mode only. Produces the abbreviation $\max$, which functions as a large operator, with limits in display style.

\medbreak

Page 52, 77. Causes a conditional vertical skip by \medskipamount , and marks the place as a fairly good one for a page break. If the \medbreak was preceded by another skip, the lesser of the two is canceled; in particular, two consecutive \medbreak s have the same effect as one.

medium space

An amount of space controlled by \medmuskip (see next entry), and automatically placed around binary operators (see that entry). It can also be requested explicitly with \> .

\medmuskip

Page 182. One of TₑX's math glue variables, set by plain TₑX with \medmuskip=4pt plus 2pt minus 4pt , or this much: | |. Generally called a medium space (see preceding entry).

\medskip

Page 40. Causes a vertical skip by \medskipamount . For examples, see pages 12, 45, 49–50, 52.

\medskipamount

One of TₑX's glue variables: it controls the amount of a \medskip . Plain TₑX sets it to 6pt plus 2pt minus 2pt . See also \eightpoint .

memory

Pages 6, 85.

\message

Prints the material that follows in braces on the screen and in the log file. See examples under \chapter , \section and \subsection .

\meti

Pages 50, 63. Starts a new paragraph and places its argument in the paragrph indentation, separated from the following text by half an em.

metric

files, see pages 6, 9, \fontdimen ; system, see page 39, units.

microcomputers

Page 7.

\mid

Page 134. Math mode only. Produces the relation $\mid$. The same symbol is obtained with | , but the spacing is different:

$K=\{(x,y)\mid x<|y|\}$ .... $K = \{(x, y) \mid x < |y|\}$

\midinsert

Page 70. Basically, TₑX sets the material between \midinsert and \endinsert at the current spot on the page or at the top of one of the following pages, depending on space constraints. More precisely, TₑX sets the material in a box and measures its height plus depth. If there is room for it on the current page, separated from the existing text by a \bigskip , TₑX prints the box there. Otherwise, TₑX treats the material as a floating insertion, as if it had been found between \topinsert and \endinsert .

Occasionally it happens that a `\midinsert` is just over the amount of space left on the current page. In such a case the resulting floating insertion may actually fit on the page, because of the elasticity of the glue, with the unfortunate result that the material appears at the top of the current page, *before* the point in the text where it is referred to. The only way to fix this (short of changing the definition of `\midinsert`) is to jiggle the contents of the insertion so the coincidence no longer happens: some 6 pt added or subtracted from its height are generally enough to do the trick.

`\midinsert` won't work inside a vertical box, or inside a horizontal box that is part of a paragraph or box of any kind; the insertion will disappear. See insertions for details.

millimeter	See `mm`.
`\min`	Pages 140. Math mode only. Produces the abbreviation min, which functions as a large operator, with limits in display style.
minus	Pages 40, 44–45, 181, 194. Keyword that introduces the shrinkability of glue.
minus sign	Pages 19, 100, 182. See also `-`.
missing	braces, page 26; control sequence, pages 171, 175; dollar sign, pages 16, 171; rule dimensions, pages 94, 115, 126.
`\mit`	Pages 33, 136. Switches to a math italic font; its effect is felt only in math mode. Mostly used for italic Greek capitals.
`\mkern`	Page 138, 140, 160, 182. Math mode only. Introduces a horizontal kern of a specified dimension, expressed in math units (see `mu`), and so dependent on the current math font and style. For this reason it is to be preferred to `\kern`, which also works in math mode. `\mkern` is followed by a dimension, not by glue. There is no corresponding vertical kerning command: vertical skips don't make sense in math mode.
`mm`	Pages 10, 13, 39. A keyword for millimeter, one of TeX's units.
mode changes	Pages 24, 26, 52, 80–83, 94. Here is a summary of the situations when TeX changes mode:

The following commands, when encountered in vertical mode, cause TeX to go into ordinary horizontal mode, starting a paragraph: any character whose category code is 11 or 12 (letter or ordinary); `\char`; any control sequence defined with `\chardef`; the unboxing commands `\unhbox` and `\unhcopy`; `\valign`; `\vrule`; `\hskip`; the springs `\hfil`, `\hfill`, `\hss` and `\hfilneg`; `\accent`; `\discretionary`; `\-`; `\ `; and any begin/end math character (normally `$`).

In ordinary horizontal mode, `\par` (or a blank line) causes TeX to wrap up and typeset the current paragraph. In restricted horizontal mode, `\par` is ignored. The following commands are incompatible with horizontal mode, so their appearance causes TeX to generate a `\par` (if in ordinary horizontal mode) or issue an error message (if in restricted horizontal mode): the unboxing commands `\unvbox` and `\unvcopy`; `\halign`; `\hrule`; `\vskip`; the springs `\vfil`, `\vfill`, `\vss` and `\vfilneg`; `\end`; and `\dump`. The same commands are forbidden in math mode, except that `\halign` is allowed in display math if it's all by itself.

In ordinary horizontal mode, a begin/end math character (normally `$`) causes TeX to look at the next token, without expanding it. If it is another similar character, TeX goes into display math mode, otherwise into ordinary math mode. Math mode ends when a matching `$` is found, and the mode reverts to what it was. In restricted horizontal mode, `$` has the same effect, but `$$` is not recognized as introducing display math mode: it creates an empty formula instead.

In any mode, the appearance of \hbox or \halign puts T_EX in restricted horizontal mode, which lasts till the end of the group delimited by braces after the command; at the end of the group the mode reverts to whatever it was before. Similarly, \vbox , \vtop , \vcenter and \valign put T_EX in internal vertical mode till the end of the corresponding group. But \vcenter is allowed only in math mode.

\models — Page 134. Math mode only. Produces the relation $\models$.

Monotype — Pages 1–2.

Montesquieu — Page 108.

\month, \monthname — Page 185. \month is one of T_EX's integer variables: it contains the current month (according to your computer's operating system). To print it numerically you must preceded by \the or \number (page 180). To print the name of the current month, say \monthname , after having copied its definition, since it is not part of plain T_EX.

Montlhéry — Page 60.

mouse — Page 7.

mouth — See lexical analysis.

\moveleft, \moveright — Pages 88, 97, 181. The construction

$$\texttt{\char`\\moveleft}\ \textit{dimension box command}$$

has the same effect as *box command*, except that it must occur in vertical mode, and the box is moved horizontally so its reference point is displaced by *dimension* relative to the enclosing box's reference point, instead of being vertically aligned with it. If *dimension* is positive, the box is moved left; if negative, the box is moved right. The *box command* can be built on the fly, with \hbox , \vbox , etc., or it can be fetched from memory, with \box or \copy . (But the unboxing commands don't make sense here: the box has to be intact.) Unsurprisingly, \moveleft is like \moveright , but the direction of motion is reversed.

\mp — Pages 133. Math mode only. Produces the binary operator $\mp$.

msam and msbm fonts — Page 34. Fonts distributed by the AMS, containing blackboard bold characters and many mathematical symbols unavailable in plain T_EX. The easiest way to use them in plain T_EX is to \input the macro files amssym.def and amssym.tex , which come with the fonts, and where control sequences for the symbols are defined; for example, \varnothing stands for $\varnothing$.

\mskip — Pages 138, 140, 182. Math mode only. Introduces horizontal glue of a specified dimension, expressed in math units (see mu), and so dependent on the current math font and style. For this reason it is to be preferred to \hskip , which also works in math mode. There is no corresponding vertical glue command: vertical skips don't make sense in math mode.

mu — Pages 140, 160. A keyword for math unit, a unit of distance equal to 1/18 of an em in the current font (which depends on the style).

\mu — Page 132. Math mode only. Produces the Greek letter μ.

\multiply — Page 183. Multiplies the contents of an integer register by an integer: \multiply\pageno by2 . To multipy a dimension register by any number—not necessarily an integer—you can say, for example, \dimen0=2.5\dimen0 , according to the general rules explained under dimensions. This doesn't work for integer registers.

\multispan   Pages 110, 114, 116, 121. Used at the beginning of an \halign (or \valign) entry, immediately after an & , \cr or \noalign construction. It is followed by an integer n, which must be in braces if it has more than one digit, and causes the entry which it introduces to span the next n columns of its row (or the next n rows of its column, for a \valign). The templates for all these columns are ignored. The use of \multispan anywhere else is an error. For the definition of \multispan , see \loop .

\multitex   Page 51. Not part of plain TEX. Makes a TEX pattern repeated horizontally: see \leaders .

Murphy's law Page 29.

\muskip   Pages 178, 182–183. Refers to one of TEX's 256 numbered registers for math glue. Except for \muskip0 through \muskip9 and \muskip255 , which can be used for temporary storage, all other registers should be allocated using \newmuskip .

\muskipdef   Makes the following control sequence an abbreviation for a \muskip construction. Used like \countdef . In practice \muskipdef is almost never used, because \muskip registers should be allocated with \newmuskip : see the previous entry and page 180.

\myfootnote   Page 62. Not part of plain TEX. A more sophisticated command than \footnote . It takes only one argument, the text, since the mark is a number incremented automatically. It can be adapted to set the note in a different font from the text. \footnote won't work inside a vertical box, or inside a horizontal box that is part of a paragraph or box of any kind; the footnote will disappear. See insertions for details.

In one respect \myfootnote is less sophisticated than \footnote : it doesn't read the text of the note as an argument, so the characters' category codes are not frozen before TEX has a chance to look at them. This is important, for example, if you're dealing with verbatim text (page 174). Here's a rewriting of \myfootnote that avoids this problem; to understand how it works, see \aftergroup :

```
\def\pre{\unskip\footnote{$^{\the\notenumber}$}}
\def\post{\global\advance\notenumber by 1}
\def\myfootnote{\pre\bgroup\aftergroup\post\let\dummy=}
```

\nabla   Page 132. Math mode only. Produces the symbol ∇.

naming   a box, page 84; a font family, page 33; a register, page 179.

\narrow   Page 167. Not part of plain TEX. The material between \narrow and \endnarrow is typeset with \leftskip and \rightskip equal to \parindent . See the following entry.

\narrower   Pages 23, 57. Causes the subsequent material to be typeset with left and right margins pushed in by the paragraph indentation: more precisely, it increases \leftskip and \rightskip by \parindent . Useful in quotations, etc. The effect of \narrower can be limited by grouping. In that case the last paragraph to which the change applies should end before the group where it started.

\natural   Page 132. Math mode only. Produces the symbol $\natural$.

natural component
              Pages 40–41, 178; see also glue.

\ne   Page 134. Math mode only. Produces the relation $\neq$, also obtained with \neq .

\nearrow   Page 134. Math mode only. Produces the relation $\nearrow$.

negations   Page 134. In plain TEX the negation of a relation is obtained by overstriking the relation with \not . The only negation that deserves a more elaborate symbol is \notin . The AMS distributes fonts that have proper negated relations: see msam and msbm fonts.

negative   dimensions, see pages 79, 93, dimensions; glue, pages 39, 44, 48, 61, 123, 159; indentation, pages 53, 55; numbers, see integers, decimal constants; \pageno , page 65; penalties, pages 76–77; spacing, pages 39, 44, 48, 61, 123, 159; springs, page 48; width, page 108.

\negspring   Page 48. Not part of plain TEX. A "negative spring" whose natural width is zero but can shrink to $-\infty$.

\negthinspace
    Pages 44, 138. Produces a negative thin space, that is, brings the adjacent symbols closer together by one-sixth of an em. Its use is rare, since in math mode \! is to be preferred.

\neq   Page 134. Math mode only. Produces the relation $\neq$, also obtained with \ne .

nested   conditionals, page 185; groups, page 22; macro definitions, page 169; quotes, see quotes.

network   Page 4.

\newbox   Pages 84, 179. Assigns a name to the number of a box register that is guaranteed not to have been used elsewhere, as long as the discipline explained on page 179 is followed. Saying \newbox\toto makes \toto synonymous with the appropriate box number, not with the box itself: you refer to the box by saying \setbox\toto , \box\toto , and so on.

    This and all other plain TEX macros starting with \new... are \outer , that is, they are not allowed to appear in macro definitions and in certain other situations. To get around this, see \outer .

\newcount, \newdimen
    Pages 62, 179–180. Assigns a name to an integer or dimension register that is guaranteed not to have been used elsewhere, as long as the discipline explained on page 179 is followed. Saying \newcount\toto makes \toto synonymous with \count n, for some (irrelevant) value of n; and similarly with \newdimen . See also \newbox above (last paragraph) for an important warning.

\newfam   Pages 33–34, 84. Assigns a name to a family of math fonts that is guaranteed not to have been used elsewhere, as long as all new families are so assigned. Saying \newfam\toto makes \toto synonymous with the appropriate family number, and \fam\toto sets the current math family to the family referred to by \toto . See also \newbox above (last paragraph) for an important warning.

\newif   Pages 65, 185. Saying \newif\iftoto defines three new control sequences: \iftoto , \tototrue and \totofalse . The last two make \iftoto equivalent to \iftrue and \iffalse , respectively. The name of the control sequence following \newif must start with if . See \iffalse to find out how \newif works. See also \newbox above (last paragraph) for an important warning.

\newmuskip, \newskip
    Page 179. Assigns a name to a math glue or regular glue register that is guaranteed not to have been used elsewhere, as long as the discipline explained on page 179 is followed. Saying \newskip\toto makes \toto synonymous with \skip n, for some (irrelevant) value of n; and similarly with \newmuskip . See also \newbox above (last paragraph) for an important warning.

newsletter   Pages 3, 7.

\newtoks   Pages 65, 179. Assigns a name to a token list register that is guaranteed not to have been used elsewhere, as long as the discipline explained on page 179 is followed. Saying, for example, \newtoks\toto makes \toto stand for with \toks n, for some (irrelevant) value of n. See also \newbox above (last paragraph) for an important warning.

Newton–Girard formulas
> Page 156.

\ni   Page 134. Math mode only. Produces the relation $\ni$, a synonym for \owns. Used by aficionados of machine language and other monstrosities:

$$\mathbf{R}^{n} \ni x \xmapsto{f} x + \varphi(x)y \in \mathbf{R}^{m}$$

\noalign   Pages 108, 120. Used between rows of an \halign, or columns of a \valign, immediately after a \cr or another \noalign. It inserts the material that follows in braces, generally a glue command, rule or box, between the rows or columns of the alignment. The material is processed in vertical mode in the case of \halign, and in horizontal mode in the case of \valign. Page 109 shows how to avoid a common mistake; pages 112 and following how to open up tables, pages 157, 159, 186 how to open up displayed equations and matrices. See also \eqalignno for how to add text between aligned equations.

\nobreak   Page 76. An abbreviation for \penalty 10000. It prevents TeX from breaking the line (if in horizontal mode) or page (if in vertical mode) at this point. But it doesn't act retroactively, so if \nobreak is preceded by glue its effect will be nullified. This causes no end of sorrow, because the preceding glue may have been put there by a macro, without your knowledge. For an example of what to watch out for, see \section and \subsection.

\noexpand   Inhibits the expansion of the next token. For an example of use, see \outer.

\noindent   Pages 12, 46, 52, 57, 72, 92. Encountered in vertical mode, it starts a paragraph without indentation. In horizontal mode it does nothing.

\nointerlineskip
> Page 93. Placed between two boxes in vertical mode, it suppresses the automatic spacing that would otherwise separate them. See also \offinterlineskip.

\nolimits   Pages 142. Math mode only. After a large operator (page 133), declares that limits (or subscripts and superscripts) should be placed in their normal positions, rather than above and below the operator, in all styles. By default, such operators get limits above and below in display style. See also \limits, \displaylimits.

\nonfrenchspacing
> Tells TeX to follow the conventions of Anglo-Saxon typesetting regarding spacing after punctuation: see \frenchspacing. This is the normal state of affairs in plain TeX.

non-letter   Page 164.

\nonstopmode
> Causes TeX to run without stopping for errors, no matter how serious, but to print error messages on your screen. Typing r in response to an error has the same effect. See also error checking and recovery.

\nopagenumbers
> Pages 64–65. Suppresses page numbering.

\normalbaselines, \normalbaselineskip
> Pages 159–160, 163, 186. \normalbaselines makes \baselineskip, \lineskip and \lineskiplimit revert to their default values \normalbaselineskip, \normallineskip and \normallineskiplimit, which are set by plain TeX to 12 pt, 1 pt and 0 pt, respectively. The idea is that these values are more or less permanent, while the values of \baselineskip, \lineskip and \lineskiplimit are changed for many purposes (as by \openup). This is used by certain macros like \matrix to ensure consistent spacing.

But the "permanent" values can also be changed, and indeed it is essential to do so to get good results with different size type: see \eightpoint .

\normalbottom

Cancels the effect of \raggedbottom , that is, makes all pages have the same height. This is the standard setting in plain TEX.

\normallineskip, \normallineskiplimit

Pages 159–160, 163, 186. See \normalbaselines .

\not

Page 134. Math mode only. Produces the symbol $/$, whose dimensions are such that it overstrikes the following symbol:

$x\not=y$, $U\not\subset V$ .... $x \neq y, U \not\subset V$

See also negations.

\notenumber Page 62. Not part of plain TEX. Integer register used by \myfootnote to keep track of note numbers.

\notin

Page 134. Math mode only. Produces the relation $\notin$. An alternative for \not\in , which gives $\notin$.

\nu

Page 132. Math mode only. Produces the Greek letter ν.

NULL

Page 172. The character with ASCII code 0, assigned by plain TEX category 9 (ignored).

\null

Pages 46, 51, 69, 90. An abbreviation for \hbox{} . Useful in preventing preceding or following glue from disappearing.

\number

Page 181. Generates the decimal representation of the following integer (see integers). The tokens generated in this way may be printed or reinterpreted as a number, depending on the context; for example, \number"10 prints 16, but \pageno=\number\pageno 0 multiplies the page number by 10!

numbering equations

Pages 155–156, 189.

\nwarrow

Page 134. Math mode only. Produces the relation $\nwarrow$.

\o, \O

Page 19. Produce the letter ø, Ø. Compare \emptyset $\emptyset$.

\oalign

Page 151. Makes a tiny one-column alignment with very little space separating rows. Used to place things above and below a character, closer than \atop will let you. Plain TEX's dot-under accent is defined with

$$\def\d#1{\oalign{#1\cr\hidewidth.\hidewidth}}$$

\obeylines

Pages 53, 57, 174–175. Causes TEX to interpret a CR in the input file as a \par command, so each line of input starts a new line of output. But a blank line in the input doesn't cause a blank line in the output! By changing the meaning of \par you can achieve special effects: see under \let for an example.

There is no command to counteract \obeylines : it remains in effect till the end of the group where it was found.

\obeyspaces

Pages 25, 174, 176. Causes TEX to interpret a SP in the input file as a \space command, so consecutive spaces are no longer merged as usual. Watch out: spaces just before the beginning of a paragraph (or at the beginning of a line, if you've said \obeylines) are still ignored, since they're read in vertical mode. To fix this, you must redefine the active space by saying {\obeyspaces\gdef {\leavevmode\space}} .

octal numbers See page 36, integers.

\oddpagefoot, \oddpagehead

Pages 65–66, 71, 183. Not part of plain TeX. In the `fancy` format, these variables contain the material that forms the footer and header of each odd-numbered page. For more details, see \evenpagefoot .

\odot   Pages 133. Math mode only. Produces the binary operator ⊙. Compare \bigodot ⊙.

\oe, \OE   Page 19, 38. Produce the letter œ, Œ.

\of   Page 145. Math mode only. Delimits the index of a radical: \root...\of{...} .

\offinterlineskip

Pages 93, 97–98, 114, 116. Eliminates altogether the space TeX automatically places between boxes and lines. Essential in tables that have vertical rules.

\oint   Page 133. Math mode only. Produces the large operator ∮, ∮ . Unlike most large operators, \oint has its limits placed in the subscript/superscript position, rather than above and below it. You can change this behavior with \displaylimits .

\oldstyle   Pages 33, 35–36, 136. Switches to a font that has old-style digits 1234567890. Works in all modes. The use of non-digits while \oldstyle is in effect gives weird results, because everything is taken from the math italic font:

\oldstyle Where is H\'el\'ene? .... *WhereisHelΘne⋆*

\omega, \Omega

Page 132, 151. Math mode only. Produce the Greek letter ω, Ω.

\ominus   Pages 133. Math mode only. Produces the binary operator ⊖.

\omit   Pages 107, 112, 114–115. Used at the beginning of an \halign (or \valign) entry, immediately after an & , \cr or \noalign construction, it causes the template for the corresponding entry to be ignored. The use of \omit anywhere else is an error. See also \loop .

open quotes   Page 18.

opening delimiter

Page 131, 135; see also \bigl , \left . No space is placed between an opening delimiter and the following symbol, but spacing is placed before an opening delimiter that is preceded by a binary operator, a relation, a punctuation mark or an inner subformula, according to the rules for each of these classes. See also \mathopen .

\openup   Pages 113, 116, 152, 154–155, 157, 159, 186. Saying \openup x, where x is a dimension, is the same as increasing \baselineskip, \lineskip and \lineskiplimit by x. The net result is that the glue placed automatically between lines and boxes that are being stacked is increased uniformly. The effect of several \openup s is cumulative. If you want to separate all the rows of an \halign (or one of the macros that perform alignments), you must place \openup before, not inside, the \halign .

The interesting thing about \openup is that, although it is a macro, its argument doesn't have to be placed in braces; TeX expects to see a dimension after \openup and will scan ahead until finding one. The way that's achieved is clever: see \afterassignment .

operating system

See implementation dependencies.

optional arguments

See \ifnextchar .

options file   Page 14.

\oplus   Pages 133. Math mode only. Produces the binary operator ⊕. Compare \bigoplus ⨁.

\or   Page 185. Separates the cases in an \ifcase. See \today and \magstep for examples.

orchestra   Page 3.

ordinary character
                    This expression has two senses. It means an input character of category 12 (page 172), that is, one that cannot appear in multi-character control sequences, but doesn't have a special meaning. But in math mode, it also means an output character, or symbol, of class 0 (page 132). Ordinary math symbols and variable-family symbols (of class 7) are not surrounded by space: $3abc$ gives $3abc$. Any space around such characters comes from an adjacent large operator, binary operator, relation, punctuation mark or inner subformula, as explained under those entries. See also \mathord.

ordinary   horizontal mode, pages 24, 80; vertical mode, page 25, 79, 105.

\oslash   Pages 133. Math mode only. Produces the binary operator ⊘.

\otimes   Pages 133, 161. Math mode only. Produces the binary operator ⊗. Compare \bigotimes ⊗.

outer level   Page 25. Something is said to be at the outer level if it isn't inside any boxes, but is being contributed directly to the page. The distinction is important because alignments, or anything else, inside boxes will not be broken across pages, but no such restriction applies to things at the outer level: see page 105 and \everycr. See also insertions and \mark.

\outer   Page 171. Placed before the definition of a macro, declares that the macro is not allowed to appear in macro arguments or replacement text, nor in the portion of a conditional that is being skipped over, nor in the preamble of an alignment. Therefore an \outer macro is treated even more strictly than \par, which is only forbidden inside macro arguments.

                    In plain TEX the following macros are declared \outer: \beginsection, \proclaim, \+, and all macros starting with \new. But often one wants to have \newcount, for example, as part of the replacement text of a macro, or inside a conditional. How to get around its outerness? One way is to use instead a non-outer macro \mynewcount which expands to \newcount. Defining \mynewcount is not trivial, because we can't just say \def\mynewcount{\newcount}, because we run into the outerness problem again! We must instead use a subterfuge:

                    $$\edef\mynewcount{\noexpand\newcount}$$

                    The idea is that \edef expands the stuff in braces before assigning it to \mynewcount — but the "expansion" of \noexpand\newcount is just \newcount! In some sense, \edef and \noexpand cancel each other, but the net effect is to allow an outer sequence to be part of a replacement text.

outline fonts   Page 31.

output   See dvi file, log file.

\over   Pages 148. Math mode only. Makes a fraction: $1\over 2$ gives $\frac{1}{2}$. The numerator and denominator form groups, and extend to the limits of the smallest enclosing group. They are typeset in a smaller style than the style in effect when the fraction is encountered: pages 139–140. They are normally centered, but that can be changed by the use of \hfill: page 33. See also the related entries \overwithdelims, \atop, \atopwithdelims, \above and \abovewithdelims.

There is also a completely different use of \over, with the \buildrel control sequence: page 170.

\overbrace   Page 145. Math mode only. Places horizontal braces above its argument, which is also read in math mode. Unlike \downbracefill, whose results are similar, \overbrace doesn't have to be put in a separate box and stacked above the main text: it takes care of everything.

overfull boxes and lines
              See pages 15–16, 41, 76, 85, 89, and \- .

\overfullrule
              Pages 16, 41. One of TEX's dimension variables: it controls the thickness of the vertical stroke ("black box") that TEX places next to an overfull box or line, to catch your eye. Plain TEX sets \overfullrule=5pt; if you say \overfullrule=0pt, TEX will not print the stroke.

\overleftarrow, \overline, \overrightarrow
              Page 145; for \overline, see also pages 150, 167. Math mode only. These macros draw arrows and bars above their arguments. The results with single letters, especially capitals, is not ideal: $\overline M$ gives $\overline{M}$. Sometimes it helps to introduce a negative thin space on the left, but that wrecks the spacing:

$\overline{\!M}+\overline{\!C}+\overline{\!Y}$ .... $\overline{M} + \overline{C} + \overline{Y}$

Another alternative is to use \bar and \vec instead, although these produces a much smaller accent.

\overwithdelims
              Page 142. Math mode only. Like \over, but encloses the fraction it between specified delimiters:

$$x+{y+z\overwithdelims[] v+w}$$ .... $x + \left[\dfrac{y+z}{v+w} \right]$

\owns   Page 134. Math mode only. Produces the relation $\ni$, a synonym for \ni .

\P   Produces the symbol ¶, in text or in math mode. In math mode it does not change size in subscripts and superscripts.

package   Pages 5, 6; see also format file.

page   blank, pages 69; breaks, pages 41, 44, 75–77; building, pages 25, 79; description language, pages 2–3, 5, 9; layout, pages 7, 40, 45, 64–77; number, pages 64, 66, 67; shipout, page 79; width page 41.

\pageno   Page 65, 181, 183–184. An integer register that contains the page number: the same as \count0. For its use, see \folio and \count .

papyri   Page 176.

\par   Pages 26, 28, 45, 47, 51–53, 75, 171, 175. Terminates the current paragraph, if TEX is in ordinary horizontal mode. In vertical mode, it is ignored. In restricted horizontal mode, it doesn't make sense (since TEX is composing a single line), and is also ignored. In math mode it causes an error.

              A blank line is normally equivalent to \par: see CR. After \obeylines, a single CR is equivalent to \par . In that case you can get interesting effects by changing the meaning of \par: see \let .

paragraphs   beginning, see mode changes; and boxes, page 82; and display math, page 131; ending, see mode changes; indentation, page 23; spacing between, page 45.

\parallel   Page 134. Math mode only. Produces the relation ∥. The same symbol is obtained with \ | , but in the latter case it is not surrounded by space.

parentheses   Pages 100, 135; see also (.

\parfillskip

Pages 47, 56, 70. One of T_EX's glue variables: it controls the glue added automatically to the end of a paragraph. Plain T_EX sets it with \parfillskip=0pt plus 1fil , so it corresponds to a weak spring, to fill up the last line. A longish paragraph, like this one, can be made to end flush with the right margin with the command \parfillskip=0pt , without the interword spacing stretching overmuch. See also \qed and \raggedleft .

\parindent   Pages 13–14, 23, 47, 49–50, 53, 55, 57, 62, 99. One of T_EX's dimension variables: it controls the automatic indentation at the beginning of a paragraph. It is also used by other macros such as \narrower and \item , so if your format, like the one used in this book, doesn't use paragraph indentation, you have to reset \parindent temporarily in order to make use of those macros.

Paris   Pages 60, 86.

\parshape   Pages 61. A command to make paragraphs with spectacular shapes. Its syntax is baggy and won't be repeated here. Applies only to the current paragraph. See also \hangafter and \displaywidth .

\parskip   Pages 13, 23, 45. One of T_EX's glue variables: it controls the glue added just before a new paragraph starts. Plain T_EX sets it with \parskip=0pt plus 1pt , adding a bit of stretchability to the page.

Pascal   Blaise, page 92; programming language, pages 6, 65; triangle, pages 123, 126.

\partial   Page 132. Math mode only. Produces the symbol ∂.

pc   Pages 32, 36. A keyword for pica, one of the commonest units in typesetting. One inch equals approximately 6 picas; one pica equals 12 points.

PC   Pages 5, 7.

\penalty   Pages 75–77, 173. Followed by an integer, instructs T_EX to consider the current spot good or bad for a line break (in horizontal mode) or page break (in vertical mode). Other things being equal, the more positive the penalty the less likely a break is to occur, and the more negative, the more likely. But many other factors influence the choice of breaks. There are only two values that are absolute: \penalty-10000 always causes a break, and \penalty10000 always prevents a break at any glue that follows—but it has no effect on preceding glue (see \nobreak). Two consecutive penalties are equivalent to the lowest. There are many abbreviations for various useful penalties: pages 76–77. Note: \penalty is not a variable, and you can't assign to it, so don't try an = between \penalty and the number.

Pensées   Page 92.

percent sign   See % , \% .

Perfect Table   Page 115.

perpetual motion

Page 174.

\perp   Page 134. Math mode only. Produces the relation ⊥.

\phantom   Page 151. Puts its argument inside a horizontal box, measures it, then typesets an empty box of same dimensions. The material in the argument does not appear on the page. See also \hphantom , \vphantom , \smash .

\phi, \Phi   Page 132. Math mode only. Produce the Greek letter ϕ, Φ. Compare \varphi φ.

phototypesetting
              Pages 1, 5.

physicists   Page 19.

\pi, \Pi   Page 132. Math mode only. Produce the Greek letter π, Π. Compare \prod $\prod$, \varpi ϖ.

pica   See pc .

PicTEX   Page 163.

pixel file   Page 9.

plain TEX   Page 5. A basic TEX format, contained in the file plain.tex . and described in detail in this book. Often TEX is used to refer to plain TEX, although in this book we tried to keep them separate.

Plato   Page 151.

Playback   Page 66.

plus   Pages 40, 44–45, 180. Keyword that introduces the stretchability of glue.

\pm   Pages 133. Math mode only. Produces the binary operator $\pm$.

\pmatrix   Pages 12, 157, 159. Creates a matrix surrounded by parentheses. Used exactly in the same way as \matrix .

\pmod   Math mode only. Prints a "modulo condition:" $$x\equiv y\pmod n$$ gives

$$x \equiv y \pmod n.$$

              Compare \bmod ; there is no \mod .

point   See . , pt .

Polish L   See \l .

portability   Pages 3–4.

PostScript   commands, page 7; fonts, pages 6, 9, 27, 29, 31, 34, 36; images, 7.

\Pr   Page 140. Math mode only. Produces the abbreviation Pr, which functions as a large operator, with limits in display style.

preamble   Pages 103, 119, 169; see also \outer .

\prec, \preceq
              Page 134. Math mode only. Produce the relations $\prec$, $\preceq$.

preloaded fonts
              Page 27.

preprints   Page 68.

\pretolerance
              One of TEX's integer variables, set by plain TEX to 100. In attempting to break a paragraph into lines, TEX starts by trying not to hyphenate any words. If there is a solution whose badness is less than \pretolerance , the best such solution is chosen. Otherwise TEX tries again, harder: see \tolerance .

              If you set \pretolerance=10000 , TEX never gets to the second stage, and no words will be hyphenated; but the spacing may be pretty bad.

\qed   Pages 71, 75. Not part of plain TEX. Creates an end-of-proof symbol □. The following macro is a good example of the use of \parfillskip: it places the same symbol flush right on the line, separated by .75 em from the text, or on a line by itself if the last line is too long.   □

```
\def\flushqed{\unskip\nobreak
  \hfil\penalty50\hskip.75em\null\nobreak\hfil\qed
  {\parfillskip=0pt \finalhyphendemerits=0 \par}}
```

Cryptic? Perhaps, but don't despair. To understand it you just need to assimilate TEX's process for breaking paragraphs into lines. See chapter 14 of *The TEXbook*, especially page 106.

\qquad, \quad

Page 44. Leave horizontal kerns, or unbreakable spaces, respectively two ems and one em wide. Both commands are very common: see examples on pages 12, 39, 44, 90, 92, 105, 116, 118, 123, 138, 153, 165. Compare \enspace.

quality of typesetting

Pages 2–3.

quotations   Page 18, 57.

quotes   Page 18. For best results, consecutive single and double quotes should be coded '{}'', ''\thinspace', '\thinspace'' and ''{}', as the case may be.

Quotation marks in the input have a special meaning when TEX is expecting to read a number: see integer.

r, R   When TEX has stopped because of an error, typing r or R makes it continue in non-stop mode: see \nonstopmode.

radicals   Page 144.

radio   Page 6.

\raggedbottom

Normally, all pages of a document start at a fixed distance from the top edge of the sheet, and end at the same distance from the bottom edge. But if the material is very heterogeneous, this rigid layout may require stretching the glue too much, and it may be better to relax it. The \raggedbottom command instructs TEX to allow the bottom margin to vary by up to 60 pt from page to page. It is the vertical analogue of \raggedright, whence its name.

\raggedleft Page 56. Not part of plain TEX. Causes TEX to justify lines on the right only, leaving the left margin ragged. Its effect extends to the end of the group where it was encountered. This command is much less common than the next one.

\raggedright

Pages 55, 99. Causes TEX to justify lines on the left only, leaving the right margin ragged. The line lengths can vary by as much as 2 ems. Very useful when the text is being set in a narrow column, or is otherwise hard to justify. There is no command to counteract \raggedright: its effect extends to the end of the group where it was encountered. See also \ttraggedright.

\raise   Pages 87, 92, 160, 181; see also \lfq. The opposite of \lower: raises a box when given a positive dimension, and lowers it when given a negative one. For details, see \lower.

random variables

Page 63.

\rangle   Pages 135. Math mode only. Produces right angle brackets ⟩. For details, see \langle.

\rceil Pages 135. Math mode only. Produces the symbol ⌉. For details, see \lceil .

\Re Page 132. Math mode only. Produces the symbol ℜ, for the real part of a complex number. If you prefer Re, change its definition to \def\Re{\mathop{\rm Re}} (page 143).

recipe Page 103.

recursion Page 174; see also \everypar .

reduction Page 31; see also \magnification .

reed Page 92.

reference point
Pages 78–79.

\refpoint Pages 98, 101. Not part of plain T_EX. Placed before a box, draws an arrow indication the box's reference point.

register allocation, pages 84, 179; arithmetic on, page 182; names, page 165; safe, page 180.

registration of fonts
Page 28.

relation Pages 131, 134, 146. A symbol or subformula of class 3. A binary operator gets a thick space \thickmuskip after it if it is followed by an ordinary character, a large operator, an opening delimiter or an inner subformula. Similarly, it gets a thick space before it if it is preceded by an ordinary character, a large operator, a closing delimiter or an inner subformula. But the space is not added in script and scriptscript style. See also \mathrel .

relativity theory
Page 136.

\relax Page 165. Tells T_EX to do nothing! It is nonetheless quite useful, because it puts an end to some activity that might otherwise invade your text. One example is given under * . For another, plain T_EX defines \quad as \hskip 1em\relax . Let's take away the \relax , an experiment a bit. All goes well until we type something like

> \def\quad{\hskip 1em}
> \quad Plush carpeting welcomes our guests ...

Then T_EX gives the inexplicable message

> ! Missing number, treated as zero.
> <to be read again>
>   h
> 1.2 \quad Plush
>   carpeting welcomes our guests ...

Do you see what happened? After \hskip 1em , T_EX is still on the lookout for plus and minus . The capital P doesn't deter it (see keywords), and after it reads plus it expects a dimension!

The moral of the story: when a macro definition ends with glue, curtail T_EX's zeal with a well-applied \relax . You can also use \relax at the beginning of macros, if you think they might be called while T_EX has unfinished business.

(However, after some commands even \relax is ignored, such is T_EX's determination to find a meaningful complement. For instance, \hbox\relax{...} is perfectly legal.)

\relbar, \Relbar
Page 146. Math mode only. Produce the relations − and =. The same symbols are obtained with - and = , but \relbar and \Relbar are appropriate for building up arrows.

\removelastskip
  Cancels an immediately preceding \vskip. For an application, see \section; see also \unskip.

repeated   pattern, page 51; templates, page 106.

replacement text
  Page 165.

reserving   a box, page 84; a register, page 179.

reset key   Page 15.

resolution   Pages 2–3.

restricted horizontal mode
  Pages 25, 80; see also \hbox.

\rfloor   Pages 135. Math mode only. Produces the symbol ⌋. For details, see \lfloor.

\rfq   Not part of plain TeX. Right "foreign" quotation marks: see \lfq.

\rho   Page 132. Math mode only. Produces the Greek letter ρ. Compare \varrho ϱ.

\rhook   Pages 134, 162. Produces the relation ꜚ. For details, see \lhook.

\right   Pages 90, 147, 155, 159. The inseparable companion of \left.

right delimiters
  See closing delimiters.

\rightarrow, \Rightarrow
  Page 134, 149. Math mode only. Produce the relations $\rightarrow$, $\Rightarrow$; the first is also given by \to. See also the next entry.

\rightarrowfill
  Pages 46, 117, 149, 162. A "spring" that makes an arrow pointing right. For details, see \leftarrowfill.

\rightharpoondown, \rightharpoonup, \rightleftharpoons
  Page 134. Math mode only. Produce the relations $\rightharpoondown$, $\rightharpoonup$, $\rightleftharpoons$.

\rightline   Page 68. Creates a line containing the material that follows in braces, flush right. For details, see \leftline.

\rightskip   Pages 48, 54–55, 47, 167. One of TeX's glue variables. It controls the amount of glue placed at the end of each line of a paragraph. For details, see \leftskip.

right-to-left scripts
  Page 7.

\rlap   Pages 48. Prints its argument, then backtracks to the starting position. The argument is read in horizontal mode, even if \rlap is being used in math mode. See examples on pages 67, 79, 111, 162, 189, 189.

\rm   Pages 28, 33, 35–36, 75, 135. Switches to a boldface font, in text or in math mode. For details, see \bf.

robustness   Page 6.

roller coaster   Page 83.

roman   fonts, pages 27–28; numerals, see next entry.

`\romannumeral`

Page 181. Generates the roman numeral for the following integer, in lowercase. The integer may follow explicitly, in any representation, or may come from an integer register. The roman numeral "representation" of a non-positive number is empty.

The `\Romannumeral` macro below is not part of plain T_EX, but comes in handy for automatic copyrights and the like. It's used just like `\romannumeral`, but prints the numeral in uppercase. To understand how it works, look up `\afterassignment` and `\uppercase`.

```
\def\Romannumeral{\bgroup\afterassignment\endroman\count255=}
\def\endroman{\uppercase\expandafter{\romannumeral\count255}\egroup}
```

Rome   Page 141.

`\root`

Page 144–145. Macro that makes radicals: `$\root 5\of{1+x^2}$` gives $\sqrt[5]{1+x^2}$ in text style and $\sqrt[5]{1+x^2}$ in display style. If all you need is a square root, use `\sqrt` instead.

Rostaing, Bjarne

Page 99.

Rough Guide to Paris

Page 60.

rows

of an alignment, pages 102, 119, 123; inserting material between, page 108; spacing between, see pages 112–113, `\openup`, `\normalbaselines`.

rules

Pages 46, 81, 86, 94; in alignments, pages 114, 121, 127. See `\hrule` and `\vrule` for synopses.

Rumanian   Pages 21, 37.

ruminant   Page 177.

runaway argument

Pages 16, 152, 171.

running   head, page 67; T_EX, page 9.

`\runningauthor`, `\runningtitle`

Pages 65, 71. Not part of plain T_EX. In the `fancy` format, these variables contain the author's name and the work's title, which get printed at the top of even- and odd-numbered pages, respectively. But you can store anything you like in them. This book's format, a variation of `fancy`, prints `\runningtitle` on even-numbered pages and the chapter title, stored in the `\chaptitle` variable, on odd-numbered pages: see `\chapter`.

s, S

When T_EX has stopped because of an error, typing s or S makes it continue in scroll mode: see `\scrollmode`.

`\S`

Produces the symbol §, in text or in math mode. In math mode it does not change size in subscripts and superscripts. Also, `\S5` and `\S~5` are both unsatisfactory in terms of spacing. You can use `$\S\,5$` instead, which gives § 5.

safe registers   Page 179.

Saint-Exupéry, Antoine de

Page 155.

St. Pol   Count of, page 60; de Léon, page 110.

Salmon, Tim   Page 60.

sample line   Page 124; see also preamble.

sans-serif fonts
> Pages 27, 30.

saving space   Page 70.

scaled   Page 31. Keyword used in registering a font, if the font is to be used at other than its design size: \font\bigten=cmr10 scaled \magstep1 .

scaled point   Pages 3, 178, 181. TeX's smallest unit: see sp .

scientific texts   Page 7.

scratch registers
> Pages 84, 179.

\scriptfont, \scriptscriptfont
> Pages 32, 34–35. Followed by a family number (see \fam), these control sequences refer to two of the family's fonts, called on for characters in script and scriptscript style, respectively (see following entry).

\scriptscriptstyle, \scriptstyle
> Page 139. Switch to scriptscript and script style, the two smallest styles used in math formulas. The change remains in effect till the end of the smallest enclosing group.

> TeX uses script style for subscripts and superscripts of an expression set in text or display style, and for the numerator and denominator of a fraction set in text style. It uses scriptscript style for subscripts and superscripts of an expression set in script or scriptscript style, and for the numerator and denominator of a fraction set in script or scriptscript style. For situations where \scriptstyle is appropriate, see pages 144, 162. \scriptscriptstyle is used very rarely: see \lfq . See also \textstyle , \displaystyle .

\scrollmode   Causes TeX to run without stopping for most errors, but to print error messages on your screen. Very serious errors like a file that cannot be found will still cause an interruption. Typing s in response to an error has the same effect. See also error checking and recovery.

\searrow   Page 134. Math mode only. Produces the relation $\searrow$.

\sec   Page 140. Math mode only. Produces the abbreviation sec, which functions as a large operator without limits.

\section   Page 51. Not part of plain TeX. Command used in this book to start a section: section 1.1 started with \section{The birth of \TeX}. It takes one argument, the section name: the chapter and section numbers are supplied automatically (cf. \chapter). Here is its definition:

> ```
> \newif\ifaftersection
> \def\beginsection#1{\removelastskip
>    \vskip 20pt plus 40pt \penalty-200 \vskip 0pt plus -32pt
>    \hskip-4.75pc{\sectitlefont\the\chapno.\the\sectno\ #1}
>    \global\advance\sectno by 1 \nobreak\medskip
>    \aftersectiontrue
>    \global\everypar{\global\aftersectionfalse\global\everypar{}}}}
> ```

It starts by removing any vertical spacing that might have been put there by some other macro, so as not to interfere with its own spacing. It then adds a goodly amount of space: 20 pt, stretchable to 60 pt. But most of the stretchability is canceled by a subsequent \vskip 0pt plus -32pt . The idea is that at most $60 - 32 = 28$ pt of space should be left if the new section starts on the same page; but if a page break occurs between sections, at the \penalty-200 , the blank space at the bottom of the page may be up to 60 pt.

Next \section prints the chapter and section numbers and the section title, and it increments the section number. A \nobreak comes next since we don't want a page break after the section title; notice that it has to preced the \medskip, or it has no effect. Finally, a flag is set (\aftersectiontrue) indicating for the use of other macros, that a section has just started; this flag is automatically cleared (\aftersectionfalse) as soon as TEX goes into horizontal mode (\everypar). See also \subsection.

semicolon See ; .

series of items Page 58.

\setbox Page 84, 98–99, 154, 178–179. The construction

$$\setbox \ n \ box \ command$$

stores something in memory box n. Here n is an integer between 0 and 255, stated explicitly or by means of a name (see \newbox). The *box command* can be built on the fly, with \hbox, \vbox, etc., or it can be fetched from memory, with \box or \copy. (But the unboxing commands don't make sense here: the box has to be intact.) When building a complicated arrangement of boxes, it pays to store intermediate results using \setbox, rather than nesting a number of \hbox and \vbox commands. In any case, you must use \setbox if you want to inspect or change the dimensions of a box: \hbox and \vbox build a box and use it right away, without letting you manipulate it further.

\setminus Page 133. Math mode only. Produces the binary relation \. Compare \backslash.

\settabs Page 124. Declares that the following \+ ... \cr material is not to be printed, but merely serves as a sample line.

\sevenbf, \sevenrm
Page 28. Plain TEX defines seven-point fonts primarily for use in math subscripts (page 35), but these two, boldface and roman, can also be used in text. The other plain TEX five-point fonts are \seveni (math italics) and \sevensy (math symbols).

\sfcode Followed by an integer (indicating a character's ASCII code), defines the space factor for spaces following that character. See \spacefactor.

shapes of characters, page 9; of paragraphs, see \parshape.

\sharp Page 132. Math mode only. Produces the symbol $\sharp$. Compare \#, which gives #.

shortcomings of TEX
Page 6.

shortcut for the preamble
Page 106.

\show Causes TEX to print the current meaning of the next token on the screen, and to wait for a CR to proceed. The meaning is also written into the log file.

\showbox Page 180. Followed by a number n, it writes the contents of box n in the log file. Here n is an integer between 0 and 255, stated explicitly or by means of a name (see \newbox). After \showbox, TEX waits for a CR to proceed. If \tracingonline is greater than zero, the same information is also printed on the screen. See also the following entry.

\showboxbreadth, \showboxdepth
Two of TEX's integer variables: they control the breadth and depth of the description of boxes caused by \showbox and \tracingoutput. For example, if \showboxdepth is 3 (the default), items nested more than three levels deep are not shown; if \showboxbreadth is 5 (the default), only the first five items in each box are shown.

\showhyphens
>
> Followed by one or more words in braces, displays on the screen all the hyphens that TeX's idea of the words' allowed hyphenation. It comes accompanied by an underfull box message, which you can ignore.

\showthe
>
> Page 88, 180. Followed by any object that can be examined using \the , such as a register or one of TeX's variables, this command causes TeX to print the object's contents on the screen, and to wait for a CR to proceed. The same information also goes into the log file.

shrinkability   See elasticity.

\sigma, \Sigma
>
> Page 132, 151, 158. Math mode only. Produce the Greek letter σ, Σ. For an end-of-word ς, use \varsigma . Compare also \sum $\sum$.

\signed
>
> Not part of plain TeX. A useful macro borrowed from page 106 of *The TeXbook*. It takes one argument, a "signature," and concludes a paragraph with it. If there is room on the current line, the signature goes there; otherwise, it goes on a line by itself. It is based on the same principle as the mathematician's darling \flushqed macro: see under \qed .
>
> ```
>         \def\signed#1{\unskip\nobreak\hfil\penalty 50
>           \hskip 2em\null\nobreak\hfil\sl#1
>           {\parfillskip=0pt\finalhyphendemerits=0\par}}
> ```

Signet Encyclopedia of Wine
>
> Page 58.

\sim, \simeq
>
> Pages 134. Produce the relations $\sim$ and $\simeq$. Compare \approx $\approx$ and \cong $\cong$.

\sin, \sinh Pages 130, 140, 143. Math mode only. Produce the abbreviations sin and sinh, which function as large operators without limits.

\skewchar
>
> Page 36. Followed by a font name, refers to a special character in the font that contains information about the positioning of math accents. Plain TeX sets \skewchar\teni='177 , for math italic, and \skewchar\tensy='60 , for the symbol font. You should do likewise if you define math fonts in other sizes.

\skip
>
> Pages 178, 180–183. Refers to one of TeX's 256 numbered registers for glue. Except for \skip0 through \skip9 and \skip255 , which can be used for temporary storage, all other registers should be allocated using \newskip .

\skipdef
>
> Makes the following control sequence an abbreviation for a \skip construction. Used like \countdef . In practice \skipdef is almost never used, because \skip registers should be allocated with \newskip : see previous entry and page 180.

skipping templates
>
> Pages 107, 110.

\sl
>
> Pages 28, 35–36, 135. Switches to a slanted font, in text or in math mode; but in math mode subscripts and superscripts don't work (page 34). Should normally be used inside a group, so its effect goes away when the group ends.
>
> In plain TeX \sl always switches to the text font \tensl , and to the math family \slfam . To set things up so that \sl switches to a slanted font in the current size, see \eightpoint and \tenpoint .

slanted   characters, correction for, see italic correction; fonts, pages 27–28, 30; lines, page 6. See also diagonal.

\slash
Produces a / in text or math mode, after which a line break is allowed. See also / . For the slash in \notin , see page 134.

\slfam
Page 35–36. A name for the slanted font family to be used in math mode. To select that family, say \fam\slfam . (The \sl command does this.) In plain T_EX this family only has a \textfont , so subscripts and superscripts don't work.

slots
See registers, boxes.

\smallbreak
Page 52, 77. Causes a conditional vertical skip by \smallskipamount , and marks the place as a somewhat good one for a page break. If the \smallbreak was preceded by another skip, the lesser of the two is canceled; in particular, two consecutive \smallbreak s have the same effect as one.

\smallint
Math mode only. Produces the symbol $\int$, which functions as a large operator for purposes of spacing.

\smallskip
Pages 40. Causes a vertical skip by \smallskipamount . For examples, see pages 45, 49–50, 52, 112.

\smallskipamount
One of T_EX's glue variables: it controls the amount of a \smallskip . Plain T_EX sets \smallskipamount=6pt plus 2pt minus 2pt . See also \eightpoint .

\smash
Pages 59, 91, 117, 151, 153, 162. This wonderful macro prints its argument but sets things up so that its height and depth are ignored. Therefore the material won't interfere with interline spacing or with the placement of underlines, etc. But beware: the smashed material may overlap with something else on the page!

\smile
Pages 134. Math mode only. Produces the relation $\smile$.

source
file, pages 3, 8, 15; of T_EX, page 6. See also discipline.

sp
Page 178. A keyword for scaled point, T_EX's smallest unit. One point is made up of 65536 scaled points.

SP
The basic rule on page 38 is that any number of consecutive space characters SP in the input and up to one CR are merged into one space token, or thrown out altogether if they come immediately after a control sequence made of letters.

To be precise, T_EX ignores characters of category 10 in the following situations: (a) after it has seen one such character; (b) at the beginning of a line; (c) after a control sequence made of letters, or a control sequence made of one character of category 10. In each case, T_EX continues to ignore such characters until it sees a character of some other category. It follows that only the first of a sequence of SP's turns into a space token.

It follows also that when spaces are made active (page 176), they are no longer merged (but spaces at the right end of a line are thrown out at a previous stage: see under CR above).

See also space tokens, spurious spaces, \space , and the next entry.

\SP
Pages 20, 38, 83, 138. In horizontal mode and in math mode, produces a space equal to the normal space between words. In vertical mode it does the same thing, after starting a new paragraph. Its use is necessary to get spacing after a control sequence made up of letters (but see also page 166):

\TeX eats up spaces . . . . T_EXeats up spaces
\TeX\ eats up spaces . . . . T_EX eats up spaces

The amount of space created by \ does not depend on whether or not it comes after punctuation; cf. the next two entries.

space tokens   After your input has been turned into tokens, space tokens are no longer combined, but there are still some situations where they are ignored (assuming they have the normal category 10). The most important of them are: in vertical and math mode (page 131); after \ignorespaces ; when TeX is looking for something, like a number, a dimension, or a keyword; at the beginning of an alignment entry (pages 104, 123); before the optional = in an assignment; and after fil , fill and filll .

In addition, there are cases where one space token is ignored: after a unit like pt ; after the optional = in a \let assignment; and after a number expressed explicitly in decimal, hex, octal or character notation. (In fact it is advisable to have a space in that position, to avoid surprises: see page 184.)

Confused? Just wait until you see chapters 24–26 of *The TeXbook.* . .

\space   Page 176. This macro is defined with \def\space{ } . Therefore \space\space creates two space tokens, which are not combined, since the merging of input spaces takes place at an earlier stage. In this respect \space is like \ , but there are important differences: \space does nothing in math mode or vertical mode, since space tokens are ignored in those modes; and the amount of space left by \space depends on whether or not it follows a punctuation mark.

\spacefactor

One of TeX's integer variables. It controls—indirectly—how much space TeX puts on the page when it sees a space in your input. Its normal value is 1000, but it is set to 3000 after a period or exclamation mark or question mark, to 2000 after a colon, to 1500 after a semicolon, and to 1250 after a comma. The higher the space factor at the moment when an input space is seen, the more stretchability, and the less shrinkability, the output space will have. In addition, if the \spacefactor is 2000 or more, the natural component of the space is also increased, not just its stretchability. The space factor after a box or rule is 1000. Here is what happens when you put some text in a box and stretch or compress the box with spread , in increments of 5 pt. The third line has its natural length.

"Oh, my! Here already?" I was surprised: she must have come running.
"Oh, my! Here already?" I was surprised: she must have come running.
"Oh, my! Here already?" I was surprised: she must have come running.
"Oh, my! Here already?" I was surprised: she must have come running.
"Oh, my! Here already?" I was surprised: she must have come running.
"Oh, my! Here already?" I was surprised: she must have come running.
"Oh, my! Here already?" I was surprised: she must have come running.

The numbers above are not set in stone. You can change the space factor at a particular point by assigning a value to \spacefactor . More importantly, the space factor after each input character can be controlled by means of the \sfcode command. For example, plain TeX says \sfcode`\.3000 . The definition of \frenchspacing , which gets rid of all niceties of spacing, is simple:

```
\def\frenchspacing{\sfcode`\.1000 \sfcode`\?1000 \sfcode`\!1000
    \sfcode`\:1000 \sfcode`\;1000 \sfcode`\,1000}
```

Two more rules. The space factor doesn't change after things like parentheses and quotes (such characters have \sfcode zero); in this way the effect of a period is still felt after parentheses and quotes. And finally, if a period (or any punctuation) comes right after a capital, its space factor is not taken into account, because TeX assumes it marks an abbreviation, rather than the end of a sentence. Therefore Donald E. Knuth gives Donald E. Knuth, even though a tie is not used after the abbreviation.

\spaceskip Pages 47, 55. One of T_EX's glue variables. It controls the normal spacing between words (that is, the spacing put there when the space factor is 1000: see the previous entry). A related variable, \xspaceskip, controls the additional spacing put in after a period, a question mark, etc. (that is, when the space factor is 2000 or more). These variables are only taken into account when their value is different from zero; if it is zero, T_EX uses instead corresponding quantities found in the metric file for the current font.

Here's the effect of changing \spaceskip and \xspaceskip in the example of the previous entry. Each line has its natural width, with \spaceskip ranging from 1 pt to 5 pt and \xspaceskip=.5\spaceskip.

"Oh, my! Here already?" I was surprised: she must have come running.
"Oh, my! Here already?" I was surprised: she must have come running.
"Oh, my! Here already?" I was surprised: she must have come running.
"Oh, my! Here already?" I was surprised: she must have come running.
"Oh, my! Here already?" I was surprised: she must have come running.

spacing between boxes or lines in vertical mode, see pages 82, 93, \openup, \normalbaselines; at end of paragraph, pages 45, 70; in math, see pages 4, 17, 36, 130–131, 138, ordinary symbols, large operators, binary operators, relations, opening delimiters, closing delimiters, punctuation; between paragraphs, page 45; between rows of an alignment, see pages 105, 112–113, 159, \openup, \normalbaselines; above and below rules, page 109; between words, see page 18, 39–40, 174, \frenchspacing and the previous two entries.

\spadesuit Page 132. Math mode only. Produces the character ♠.

\span Placed in lieu of an & in an alignment row, indicates that the entries before and after it should be combined, to form a single entry spanning both columns. The entries are plugged into their templates, as usual, unless \omit is used as well. The \multispan macro is based on \span: see \loop.

Spanish punctuation
       See ¡, ¿.

\special Page 7. A command that is not interpreted by T_EX, but put into the dvi file together with its argument, for the benefit of the printer driver. Allows the inclusion of figures and other tricks in an implementation-dependent way.

spectral sequence
       Page 130.

splitting a box, page 98; a document, page 13.

\splittopskip
       Page 99. One of T_EX's glue variables. It controls the amount of glue placed at the top of a box split with \vsplit. The natural component of the glue is natural component of \splittopskip minus the height of the first enclosed box; but if that would make the glue negative, no glue is put in. The elasticity of the glue is the elasticity of \splittopskip.

spread Page 89. Keyword used in fixing the dimensions of a box from the outside.

springs Pages 42, 49, 89, 104, 178, 187; and \eqalign, pages 154, 187; in macros, page 43.

springlike creatures
       Pages 46–48, 51, 120.

spurious spaces
       Pages 35, 104, 110, 168–169, 173.

\sqcap, \sqcup
        Pages 134. Math mode only. Produce the relations ⊓, ⊔.

\sqrt
        Page 144. Math mode only. Sets its argument under a square root sign. See also \root .

\sqsubseteq, \sqsupseteq
        Pages 134. Math mode only. Produce the relations ⊑, ⊒.

\ss
        Pages 19, 36. Produces the letter ß.

\star
        Pages 134. Math mode only. Produces the relation ⋆.

stack
        Pages 23, 79.

stacking up symbols
        Page 148.

staggering subscripts and superscripts
        Page 136.

standard
        for on-line documents, page 4; for TeX, page 6.

Stanford University
        Page 1.

starting
        a new page, page 69; in midpage, page 68; a paragraph, see mode change.

stopping TeX
        See \bye , \end .

storekeeper
        Page 24.

storing
        in a box, page 84; in a register, page 179.

St. Pol
        Count of, page 60; de Léon, page 110.

Strasbourg
        Pages 19, 54, 71, 110.

stretchability
        See elasticity.

\string
        Page 174. If followed by a character token, generates the same character with category 12 (ordinary). If followed by a control sequence, generates its name, also with characters of category 12.

string of beads
        Pages 24, 78–79.

strong spring
        Page 43; see also \hfill , \vfill .

structured language
        Page 128.

\strut
        Pages 112–113, 126, 151. Produces an invisible rule of width zero, height 8.5 pt and depth 3.5 pt. Used to uniformize the distance between baselines in alignments, etc. More generally, any invisible rule or box with a similar purpose is called a strut: see examples on pages 99, 120, 141, and \mathstrut .

        For efficiency, the rule is stored in a box called \strutbox , which should be modified when the \baselineskip changes: see \eightpoint .

style
        files, pages 5, 14, 50, 53; in math, pages 133, 138, 140, 144.

SUB
        Page 172. The character with ASCII code 1, assigned by plain TeX category 8 (subscript). On some keyboards this character appears as a down-arrow, which is more mnemonic than _ .

subscripts
        Pages 32, 34, 91, 135–136, 172. A subscript is introduced by any character of category 8. See also the previous entry.

\subsection   Page 51. Not part of plain T_EX. Command used in this book to start a subsection; it takes the subsection name as an one argument. Here is its definition:

$$\verb|\def\subsection#1{\ifafterretrait\else|$$
$$\verb|\medbreak\fi{\subsectitlefont#1}\nobreak\smallskip}|$$

If it occurs right after a \section command, the title is added after the spacing left by that command (at which no break is allowed). Otherwise, a \medbreak is placed before the title, indicating a fairly good place for a page break. After the title, breaks are again forbidden.

\subset, \subseteq
    Pages 134. Math mode only. Produce the relations $\subset$, $\subseteq$.

subtraction   Page 183.

\succ, \succeq
    Pages 134. Math mode only. Produce the relations $\succ$, $\succeq$.

\sum   Pages 133, 142. Math mode only. Produces the large operator $\sum$, $\sum$. Compare \Sigma Σ.

\sup   Page 140. Math mode only. Produces the abbreviation sup, which functions as a large operator, with limits in display style.

SUP   Page 172. The character with ASCII code 10, assigned by plain T_EX category 7 (superscript). On some keyboards this character appears as an up-arrow, which is more mnemonic than ˆ .

\supereject   Page 76. Forces a page break; the comments made under \eject apply. The difference between the two commands is that \supereject will also print, on subsequent pages, any floating insertions that may be held over in T_EX's memory.

superscripts   Pages 32, 34, 91, 135–136, 172. A superscript is introduced by any character of category 7. See the entry SUP above.

superscript characters
    In addition to introducing supersripts, characters of category 7 have another function: when such a character occurs twice in a row, the duo combines with the subsequent character, as explained under ˆˆ .

\supset, \supseteq
    Pages 134. Math mode only. Produce the relations $\supset$, $\supseteq$.

support environment
    Page 6.

\surd   Page 132. Math mode only. Produces the symbol $\sqrt{}$, preferred by some over \sqrt :
    $\verb|$\surd a+\surd\,(b+1)$|$ .... $\sqrt{a} + \sqrt{(b+1)}$

\swarrow   Page 134. Math mode only. Produces the relation $\swarrow$.

synonyms   Page 165; see also \let .

\system   Page 188. Not part of plain T_EX. Allows the easy coding of systems of equations.

system dependencies
    Pages 6–8, 10, 15–16, 29, 188; see also \dump , \input , CR.

\t   Page 20. Tie-after accent: T\t uut gives Tu͡ut.

TAB   Page 172. The character with ASCII code 9, assigned by plain T_EX category 10 (space). Plain T_EX also defines \TAB to be equal to \ , so the two characters are everywhere equivalent.

tab character   See alignment separator.

\tabalign   A variation of \+ that can appear inside macro arguments.

tabbing   Pages 102, 122–129; comparison with \halign , page 129; and springs, page 127.

tables   See alignments.

\tabskip   Pages 105, 117–118, 120, 186, 190. One of T_EX's glue variables. Its value while the preamble of an alignment is being read controls the spacing between the alignment's columns (or rows, for a \valign).

tall formulas   Page 142.

\tan, \tanh   Pages 130, 140, 143. Math mode only. Produce the abbreviations $\tan$ and $\tanh$, which function as large operators without limits.

\tau   Page 132. Math mode only. Produces the Greek letter τ.

technical report
            Page 68.

template   Page 103; see also \multispan , \omit , \span .

\tenbf   Pages 28, 34–35. Plain T_EX's ten-point boldface font; generally activated through \bf .

\teni, \tenit
            Pages 28, 33, 35. Plain T_EX's ten-point italic fonts; \teni is for use in math and \tenit for use in text. Generally activated through \it .

\tenpoint   Page 35. Not part of plain T_EX. Command to restore plain T_EX's font sizes across the board: after \tenpoint , an \rm brings in an ten-point roman, \it brings in ten-point italics, and so on. The listing on pages 35 handles all the fonts, but there is also the matter of interline spacing. For best results, add before the last line of the definition:

```
\abovedisplayskip=12pt plus 3pt minus 9pt
\belowdisplayskip=\abovedisplayskip
\abovedisplayshortskip=0pt plus 3pt
\belowdisplayshortskip=7pt plus 3pt minus 4pt
\smallskipamount=3pt plus 1pt minus 1pt
\medskipamount=6pt plus 2pt minus 2pt
\bigskipamount=12pt plus 4pt minus 4pt
\setbox\strutbox=\hbox{\vrule height8.5pt depth3.5pt width0pt}%
\normalbaselineskip=12pt \normalbaselines
```

\tenrm   Pages 27–28, 33, 35, 64. Plain T_EX's ten-point roman font, the granddaddy of all fonts. You get it by default when you start T_EX, and whenever you say \rm .

\tensl   Pages 28, 35. Plain T_EX's ten-point slanted font; generally activated through \sl .

tensors   Page 136.

\tensy   Page 35. Plain T_EX's ten-point math symbol font.

\tentt   Page 28, 35. Plain T_EX's ten-point typewriter font; generally activated through \tt .

Terre des Hommes
            Page 155.

\TeX   Pages 6, 38, 47, 51, 88, 96. Produces the T_EX logo.
            The T_EXbook page 1, 6, 33, 36, 51, 61, 62, 67, 89, 120, 131, 176, 181–182, 186–187, 189.

T_EXtures   Page 7; see also \dump .

T_EX Users Group

Pages 1, 5, 59. Joining the T_EX Users Group entitles you to a subscription to *TUGboat*, a journal containing news, tutorials, program listings, conference announcements, advertisements, etc. The TUG office itself is a primary source of information on T_EX problems; when the staff does not know the answer to a question, it can generally put you in touch with someone who does. The address is P.O. Box 9506, Providence, RI 02940.

text editor, pages 6–8; inside formula, pages 137–138; italic font, pages 28, 33, 135; math mode, page 25, 130; processor, pages 5, 8; style, see \textstyle .

\textfont Page 32, 34–35, 101, 139. Followed by a family number (see \fam), these control sequences refer to two of the family's fonts, called on for characters in display or text style (see \textstyle).

\textindent Pages 62. At the beginning of a paragraph, places its argument in the paragrph indentation, separated from the following text by half an em; in other words, it acts like \meti , except that it doesn't start a new paragraph.

\textstyle Page 139. Switches to text style, the style that T_EX starts in when enters ordinary math mode, also used for the numerator and denominator of a fraction set in display style. The change remains in effect till the end of the smallest enclosing group. For a situation where \textstyle is appropriate, see page 142. See also \displastyle , \scriptstyle .

tfm file Pages 9, 30.

\the Pages 65, 71, 178, 180–183. Generates an explicit representation of the following object. It can act on many different things: registers of all types (\the\count0), variables of all types (\the\parindent), other constructions that represent integers or dimensions (\the\mathcode\`: or \the\fontdimen5\font), control sequences defined with \chardef and \mathchardef (\the\alpha generates 267 , the decimal version of "010B), and even fonts (see \font). The result of a \the construction is a string of character tokens of category 12 (ordinary), with two exceptions: a font yields a single control sequence, and a token list register yields its contents, whatever they may be.

The tokens generated by \the are spliced into the input. To have them printed on the screen, use \showthe instead.

\theta, \Theta

Page 132. Math mode only. Produce the Greek letter θ, Θ. Compare \vartheta ϑ.

thick space An amount of space controlled by \thickmuskip (see next entry), and automatically placed around relations (see that entry). It can also be requested explicitly with \; .

\thickmuskip

Page 182. One of T_EX's math glue variables, set by plain T_EX with \thickmuskip=5pt plus 5pt , or this much: | |. Generally called a thick space (see preceding entry).

thin space An amount of space controlled by \thinmuskip (see entry), and automatically placed around large operators and inner subformulas, and after punctuation (see those entries). It can also be requested explicitly with \, . See also \thinspace , \negthinspace .

thinking reed Page 92.

\thinmuskip Page 182. One of T_EX's math glue variables, set by plain T_EX with \thinmuskip=3pt , or this much: | |. Generally called a thin space (see entry).

\thinspace Page 44, 138. Leaves a horizontal kern, or unbreakable space, one-sixth of an em wide. Its use is rare, since in math mode \, is to be preferred.

tie   Pages 18, 21, 39, 45, 76, 173.

tie-after accent
               Page 20.

\tilde   Pages 137. Math mode only. Places a tilde over the following character: $\tilde{a}$. Its text coun-
               terpart is \~ . See also \widetilde .

Time magazine
               Page 2.

\time   One of TEX's integer variables: it contains the time the TEX run started (according to your
               computer's operating system), expressed in minutes after midnight. To print it you must
               preceded by \the or \number (page 180).

\times   Pages 29, 133, 145, 153. Math mode only. Produces the binary operator $\times$.

Times fonts   Pages 27, 29, 34, 96.

title   page, pages 66, 68, 71; placement, page 69. See also \runningtitle .

\titlepagefalse
               Pages 67, 185. Not part of plain TEX. In the fancy format, causes the current page not
               to be treated as a title page, which means the page's header and footer are the ones stored
               in \titlepagehead and \titlepagefoot , rather than their \odd... or \even...
               counterparts.

\titlepagefoot, \titlepagehead
               Pages 65–66. Not part of plain TEX. In the fancy format, these variables contain the
               material that forms the footer and header of the title page, that is, the first page of the run.
               See also the preceding entry and \evenpagefoot .

\titlepagetrue
               Pages 67, 185. Not part of plain TEX. The opposite of \titlepagefalse .

\tm, \tmfam   Pages 34–36. Not part of plain TEX. In this book's format, \tm switches to a Times Roman
               font, and \tmfam is the family name of a Times Roman family (used for digits).

to   Pages 42, 49, 88, 96, 98, 117–118, 163, 181, 189. Keyword used to fix the size of a box
               from the outside.

\to   Pages 134. Math mode only. Produces the relation $\rightarrow$, also obtained with \rightarrow .

\today   Not part of plain TEX. Produces the date; its definition is
                           \def\today{\monthname\ \number\day, \number\year}
               (\monthname is also not part of plain TEX).

token   Pages 177–180; see also macro expansion.

token list register
               Page 178, 182; see also the next entry.

\toks   Pages 178, 182. Refers to one of TEX's 256 numbered registers for token lists. Except for
               \toks0 through \toks9 and \toks255 , which can be used for temporary storage, all
               other registers should be allocated using \newtoks .

\tolerance   One of TEX's integer variables, set by plain TEX to 200. If TEX is unable to break a paragraph
               into lines without hyphenating any words, it tries again after having determined all allowed
               spots for hyphenation. This time around it tries to limit the badness to \tolerance . If it
               fails again, it will let you know by printing some overfull lines.

Tolkien, J. R. R.
Page 11.

\top   Page 134. Math mode only. Produces the relation ⊤.

\topinsert   Page 70. T_EX sets the material between \topinsert and \endinsert at the top of the current page, if it fits on the page together with the text that is already there. If not, it is deferred till the next page. If there are many of these floating insertions on the same page, they may need several subsequent pages to be printed.

\topinsert won't work inside a vertical box, or inside a horizontal box that is part of a paragraph or box of any kind; the insertion will disappear. See insertions for details.

\topmark   The mark text that \botmark had as the preceding page was completed. See \mark.

torture test   Page 6.

Total Book of Bicycling
Page 99.

\tracingcommands, \tracingmacros
Page 26. Two of T_EX's integer variables. When the value of \tracingcommands is non-zero, T_EX registers in the log file every primitive that it sees. When \tracingmacros is non-zero, T_EX registers each macro, its arguments and its expansion. A great help in debugging, but interpreting the log requires a bit of practice. For best results, set both variables to 2,

\tracingonline
Pages 26, 180. One of T_EX's integer variables. When non-zero, causes T_EX to print on the terminal everything that goes into the log file, including the debugging information caused by the preceding and following variables. Therefore it should be used with care, or you'll get a thousand lines scrolling by on your screen.

\tracingoutput
One of T_EX's integer variables. When non-zero at the time T_EX prints a page, it causes T_EX to print on the log file the contents of the page in symbolic form.

transmission of files
Page 4.

Três Corações
Page 20.

\triangle   Page 132. Math mode only. Produces the symbol △. Compare \bigtriangleup △ and the next entry.

\triangleleft, \triangleright
Page 133. Math mode only. Produce the binary operators ◁ and ▷.

trigraph   Page 175.

triptych   Page 87.

true   Pages 32, 39. Keyword that causes a following unit to be treated as absolute, that is, not subject to \magnification. Its use is not allowed before em, ex and mu, since these units are defined in terms of the current font.

\tt   Pages 28, 35–36, 135. Switches to a typewriter font, in text or in math mode; but in math mode subscripts and superscripts don't work (page 34). Should normally be used inside a group, so its effect goes away when the group ends.

In plain TeX \tt always switches to the text font \tentt, and to the math family \ttfam. To set things up so that \tt switches to a slanted font in the current size, see \eightpoint and \tenpoint.

\ttfam   Pages 35–36. A name for the typewriter font family to be used in math mode. To select that family, say \fam\ttfam. (The \tt command does this.) In plain TeX this family only has a \textfont, so subscripts and superscripts don't work.

\ttraggedright

Spaces in typewriter fonts have no elasticity, since they're supposed to be as wide as other characters. This makes justifying lines just about impossible, so any text of more than one line is best done with ragged margins. The \ttraggedright macro calls \tt and turns on ragged-right mode.

TUGboat   Pages 1, 5–6; see also TeX Users Group.

Two Friends   Page 86.

typefaces   design of, page 2; see also fonts.

typesetting, professional
                Pages 3, 10, 24.

typewriter fonts
                Pages 27–28, 30, 122–123, 174. In TeX's typewriter fonts all characters and spaces have the same width; in addition, most of the ligatures are absent, the exceptions being ! ' and ? ', which give ¡ and ¿.

typos   Pages 10, 15, 166.

\u   Page 20 Places a breve accent above the following character: ă. Works only in text mode.

ugliness   See badness, beauty.

umlaut   See \".

unary operator
                See binary operator.

unboxing   See \unhbox, \unvbox.

unbreakable space
                See kern, tie.

undefined control sequence
                Page 14.

\underbar   Underlines its argument. Can only be used outside math mode. \underbar ignores descenders: I want you! If you don't like this, go into math mode and use \underline.

\underbrace   Page 146. Math mode only. Places horizontal braces underneath its argument, which is also read in math mode. Unlike \upbracefill, whose results are similar, \underbrace doesn't have to be put in a separate box and stacked above the main text: it takes care of everything.

underful boxes, lines, pages
                Pages 15, 41, 56, 69, 74, 89, 99, 104, 117, \showhyphens.

\underline   Page 145, 150. Math mode only. Underlines its arguments. To maintain uniformity, use \mathstrut:

$\underline q +\underline r$ .... $\underline{q} + \underline{r}$

$\underline {\strut q} +\underline {\strut r}$ .... $\underline{q} + \underline{r}$

use of this command to offer \uppercase{toto} as an alternative to TOTO ? That would be silly indeed.

The trick is to expand the material in braces before \uppercase has a chance to look at it. The way to accomplish that is the precede the left brace with \expandafter , like this: \uppercase\expandafter{\foo} . Now everything works nicely. See \romannumeral for another application.

\upsilon, \Upsilon

Page 132, 151. Math mode only. Produce the Greek letter υ, Υ. The lowercase is hard to distinguish from an italic 'v', but why is the capital so little used?

user-friendly Page 7.

using a register

Page 180.

\v   Page 20. Places a háček above the following character: \v c gives č. Works in text only; for math mode, see \check .

\vadjust   In ordinary horizontal mode, inserts the material that follows in braces between the current line and the next, without disturbing the composition of the current paragraph. For example, \vadjust{\smallskip} at this point causes the skip that you see here. Similarly, \vadjust{\vfil\eject} breaks the page after the current line, without interrupting the paragraph. For another example, see \marginnote .

\valign   Pages 119 and following. Makes "vertical alignments," that is, alignments by columns instead of by rows. It's used in the same way as \halign , but all roles are reversed: horizontal becomes vertical, rows become columns, and so on.

Vaux-le-Vicomte

Page 60.

\varepsilon, \varphi, \varpi.

Page 132. Produce the Greek letters, ε, φ, ϖ. Compare \epsilon ϵ \phi ϕ, \pi π.

variables   Pages 5, 23, 179.

variable-family characters

Page 131. In math mode, a character of class 7 is treated as if it were of class 0 (ordinary), except that its family is disregarded when \fam has a value between 0 and 15, and the character is taken instead from family \fam . For details, see \fam and ordinary characters.

\varnothing Math mode only. Not part of plain TeX. Produces the empty set symbol $\varnothing$. To use this command you must have the symbol fonts distributed by the AMS: see msam and msbm fonts.

\varr   Pages 161–162. Math mode only. Not part of plain TeX. Draws a vertical arrow with labels, to be used in diagrams.

\varrho, \varsigma, \vartheta.

Page 132. Produce the Greek letters, ϱ, ς, ϑ. Compare \rho ρ, \sigma σ, \theta θ.

\vbadness   Page 99. One of TeX's integer variables: a threshold for complaints about underfull vertical boxes. For details, see \hbadness . Plain TeX sets \vbadness=1000 .

\vbox   Page 79. Introduces a vertical box, made by stacking one above the other the elements in its interior. Its baseline coincides with the baseline of the topmost box in it. Examples of use appear on pages 22, 26, 43, 67, 88, 97, 99, 155; see also **boxes**.

\vcenter   Pages 79. Math mode only. Introduces a vertical box which, before being added next to the formula being composed, is shifted vertically so its top and bottom are equidistant from the axis. Loosely speaking, the resulting box's baseline goes through its center. Examples of use appear on pages 22, 84, 89, 96, 100, 148, 155, 162, 186; see also **boxes**.

\vdash   Page 134. Math mode only. Produces the relation ⊢. Compare \dashv ⊣.

\vdots   Pages 12, 160. Math mode only. Produces three vertical dots ⋮.

\vec   Page 137, 145. Math mode only. Places a small bar over the following character: $\vec{a}$. For a longer arrow, or an arrow over several characters, see \overrightarrow.

\vee   Pages 133. Math mode only. Produces the binary operator ∨, also obtained with \lor.

vegetables   Pages 75, 109.

Velocio   Page 99.

verbatim mode
    Pages 128, 174; see also | , \myfootnote.

versatility   Page 4.

\vert, \Vert
    Page 132. Math mode only. Produce single and double vertical bars | and ‖. Synonymous with | and \| : see those entries for details.

vertical   alignments, pages 119 and following; arrows, page 146; bars, see | and \| ; boxes, see boxes, \vbox, \vtop, \vcenter; dots, page 160; mode, see pages 24–25, 52, 79, 82, 88, 109, 120, mode change; rules, see \hrule, \vrule; spacing, see pages 40, 52, interbox and interline spacing.

\vfil   Pages 42, 45. A weak vertical spring: it stretches to fill the available space in a box, page, etc., unless there is a stronger spring (see the next entry) in the same box. Several \vfil s in the same box share the available space equitably. Although a primitive, \vfil is essentially the same as \vskip 0pt plus 1fil.

\vfill   Pages 42–43. A strong vretical spring: it stretches to fill the available space in a box, page, etc., preventing weak springs from stretching altogether. Several \vfill s in the same box share the available space equitably. Although a primitive, \vfill is essentially the same as \vskip 0pt plus 1fill. See examples of use on pages 45, 71, 120.

\vfilneg   Page 77. A weak vertical spring whose stretchability is infinitely negative: although a primitive, \vfilneg is equivalent to \vskip 0pt plus -1fil. Unlike \vss, this spring doesn't shrink to a negative length: its only use is to cancel another, positive, spring placed somewhere else in the same box. An example of use is given under \filbreak.

\vfuzz   Page 16. One of T_EX's dimension variables, set by plain T_EX to .1 pt. An overfull page or vertical box is not reported if the excess material is less than \vfuzz. See \hfuzz.

\vglue   Pages 44–45, 69–70. Horizontal "glue" that will not disappear at a page break. It's really an invisible rule followed by normal glue.

visible springs   See **leaders**.

\voffset   One of T_EX's dimension variables; it controls the vertical offset of the text with respect to your sheet of paper. When \voffset is zero, the default, your printer is supposed to place the top margin one inch from the edge. A positive offset moves the margin down, and a negative one up. Useful in centering your text when \vsize is different from its default value of 8.9 in, or your paper has height different from 11 in.

Voltaire   Pages 61, 108.

\vphantom   Pages 151. Puts its argument inside a vertical box, measures its height and depth, then
               typesets an empty box of zero width and same height and depth. The material in the argument
               does not appear on the page. See also \phantom , \hphantom , \smash .

\vrule   Pages 94–96. Creates a rule (straight line, or rectangle) of specified dimensions. The full
               construction is

$$\verb|\vrule height| \ h \ \verb|depth| \ d \ \verb|width| \ w$$

               with the attributes always in that order. If height or depth is missing, it's set to the
               height or depth of the immediately enclosing horizontal box or line; if width is missing, it
               is set to .4 pt.

               \vrule must appear in horizontal mode, or TeX will start a new paragraph: pages 52,
               81. But see the next entry for an exception. To draw a vertical rule in vertical mode, use
               \hrule : pages 97, 117, 126. For use in alignments, see pages 115, 120, 126. For other
               examples of use, see pages 90, 98.

\vrulefill   Page 120. Not part of plain TeX. In vertical mode, creates a vertical rule of width .4 pt and
               stretching up and down as far as \hfill would in the same circumstances. Its definition is
               \def\vrulefill{\leaders\vrule\vfill} , showing that inside \leaders a vertical
               rule can be used in vertical mode.

\vsize   Pages 13–14, 181. One of TeX's dimension variables: it controls the height of the page.
               Plain TeX sets \hsize=8.9in .

\vskip   Page 40. Creates vertical glue of a specified size, with specified stretchability and shrinka-
               bility. For more details, see glue.

\vsplit   Pages 98–100. The construction \vsplit *n* to *height*, where box *n* is a vertical box and
               height is any dimension, produces a vertical box obtained by skimming off the top of box
               n. The choice of a breakpoint is made exactly as if TeX were filling a page with \vsize
               equal to *height*; in particular, you can influence it with penalties, and so on.

\vss   Pages 49. Vertical spring that can stretch or shrink indefinitely, taking on a negative height.
               Useful for things that should be allowed to spill over: see page 89.

\vtop   Page 79. Introduces a vertical box, made by stacking one above the other the elements in its
               interior. Its baseline coincides with the baseline of the bottommost box in it. Examples of
               use appear on pages 22, 59, 89, 97; see also boxes.

Walton, Bill   Page 99.

\wd   Pages 88, 92, 181, 183. The construction \wd *n* gives the width of box *n*. For details, see
               \dp .

weak springs   Page 43; see also \hfil , \vfil .

\wedge   Pages 133. Math mode only. Produces the binary operator $\wedge$, also obtained with \land .

\widedotfill
               Pages 47–48. Not part of plain TeX. Produces leaders that align vertically and are slightly
               more spaced than \dotfill .

\widehat, \widetilde
               Page 137. Math mode only. Place over the following group a circumflex or a tilde that grows
               as needed (but not beyond about three characters): $\widehat{ab}$ gives $\widehat{ab}$.

widow line   Page 75. The last line of a paragraph stranded on a page by itself. You can prevent widow
               lines by setting \widowpenalty=10000 .

width of boxes, see pages 79–80, \wd ; of characters, page 9; of column, pages 103, 124, 127; of page, see \hsize ; of rules, see next entry. See also dimensions.

width Pages 94, 96. An optional specification after \hrule or \vrule ; must be followed by a dimension. A horizontal rule without an explicit width will be as wide as the immediately enclosing box, or the page; a vertical rule without an explicit width will have width .4 pt.

window Page 7.

Winston, Pat Page 2.

wizard Page 11; see also aspiring wizard.

words spacing between, see pages 40, 47, \frenchspacing , \spacefactor .

\wp Page 132. Math mode only. Produces the symbol ℘.

\wr Pages 133. Math mode only. Produces the binary operation ≀. Also useful next to an vertical arrow in diagrams: page 162.

write in Pascal: page 65.

Wroclaw Page 19.

WYSIWYG Pages 2–3, 7–8, 10, 166.

x, X Page 15. When T_EX has stopped because of an error, typing x causes it to quit the run.

\xi, \Xi Page 132. Math mode only. Produce the Greek letter ξ, Ξ.

\xspaceskip Pages 47, 55. One of T_EX's glue variables. It controls the extra spacing between words (in addition to the normal one) that is placed after a period, question mark, etc. See \spaceskip for details.

\year One of T_EX's integer variables: it contains the current year (according to your computer's operating system). To print it you must preceded by \the or \number (page 180). See also \romannumeral .

\zeta Page 132. Math mode only. Produces the Greek letter ζ.